# Lecture Notes in Physics

Edited by H. Araki, Kyoto, J. Ehlers, München, K. Hepp, Zürich
R. Kippenhahn, München, H. A. Weidenmüller, Heidelberg,
J. Wess, Karlsruhe and J. Zittartz, Köln
Managing Editor: W. Beiglböck

## 284

## D. Baeriswyl M. Droz
## A. Malaspinas P. Martinoli (Eds.)

Gwatt Workshop (10th : 1986 : Gwatt, Switzerland)

# Physics in Living Matter

Proceedings of the Tenth Gwatt Workshop
Held in Gwatt, Switzerland, October 16−18, 1986

Springer-Verlag
Berlin Heidelberg New York London Paris Tokyo

**Editors**

Dionys Baeriswyl
Institut für Theoretische Physik, ETH Zürich
Hönggerberg, CH-8093 Zürich, Switzerland

Michel Droz
Andreas Malaspinas
DPT, Université de Genève
24, quai E.-Ansermet, CH-1211 Genève 4, Switzerland

Piero Martinoli
Institut de Physique, Université de Neuchâtel
1, rue A.-L. Breguet, CH-2000 Neuchâtel, Switzerland

ISBN 3-540-18192-X Springer-Verlag Berlin Heidelberg New York
ISBN 0-387-18192-X Springer-Verlag New York Berlin Heidelberg

Printing: Druckhaus Beltz, Hemsbach/Bergstr.;
Bookbinding: J. Schäffer GmbH & Co. KG., Grünstadt
2153/3140-543210

# PREFACE

Physicists have always been fascinated by the puzzling world of
biological phenomena, but recently the temptation of applying physical
ideas and methods to living matter has increased dramatically. This is
not only a result of improved experimental techniques and data analysis
but also of a growing interest in complex structures and dynamics. The
improved ability of dealing with many degrees of freedom allows us to
study theoretically the emergence of structures and patterns on a
scale which is typically much larger than the size of the microscopic
constituents. Viscous fingering and the roughening transition are prominent examples. Topological defects, which play an important role in
solid state physics and field theory, also belong to this class. Often
these structures do not represent the true thermodynamic equilibrium;
they can grow and decay in time or even freeze out. Thus nontrivial
spatial patterns are frequently associated with interesting time evolutions.

The possible impact of physics in life sciences is twofold. On the
one hand the powerful experimental and theoretical techniques developed for studying complex physical phenomena can certainly be very
useful in the biological context.  On the other hand certain physical
concepts such as symmetry and symmetry breaking, linear and nonlinear
stability, frustration and constrained dynamics are likely to be equally useful. It was the aim of the tenth workshop in Gwatt to elucidate
this double role of physics in the  study of living matter. Since it
was obviously impossible to cover exhaustively such a wide subject we
tried to make an exemplary selection of topics.

Part I deals with the structural and functional building blocks,
the biomolecules, and their role in the evolution process. H. Frauenfelder's contribution can serve as a clear illustration of the general
theme of the workshop. Part II is devoted to symmetry and structure.
Y. Bouligand shows that symmetries observed in biological systems are
strikingly similar to those observed in certain physical systems, in
particular in liquid crystals. He also suggests that symmetry breaking is intimately connected to the emergence of life.W. Braun, U. Aebi
and P. Bösiger explain experimental techniques for investigating the
structure of proteins, cells and organs, especially nuclear magnetic

resonance and electron microscopy. It becomes clear that careful image processing is essential for extracting detailed structural information from raw data. Part III is concerned with thermodynamics and transport properties of living matter, in particular of biomembranes. O. Mouritsen demonstrates that mathematical modeling can provide new insight into the relation between structure and biological function, whereas E. Neher describes a refined experimental technique which allows one to detect electrical currents flowing through single tiny channels across membranes.

Part IV is devoted to neural networks. K. Hepp and V. Henn describe the neural pathways associated with visual perception and subsequent eye movements. H.R. Lüscher attributes the transfer of stimuli to muscles to the cooperative action of a random neural network. Models for the processes of learning, storage and retrieval of information in the central nerve system are described by R.M.J. Cotterill and W. Kinzel.

We are grateful to E. Kellenberger (Biozentrum Basel), K. Hepp (ETH Zürich) and H.R. Zeller (Brown Boveri Research Center Baden) for their advice in establishing the scientific program. The meeting was financially supported by the Swiss National Science Foundation, the Swiss Physical Society, the chemical industry in Basel (Ciba-Geigy, Hoffmann-La Roche, Lonza and Sandoz) and the research laboratories of Brown Boveri Baden, IBM Zürich and RCA Zürich. We also thank the Evangelische Heimstätte Gwatt for providing a comfortable housing which greatly facilitated fruitful conversations.

We hope that the present collection of papers together with the numerous references will help to stimulate the curiosity of physicists about biological problems. At the same time we hope that this small volume will further the goodwill of biologists towards the attempt of physicists to advance into the complex field of living matter.

Geneva, March 1987

Dionys Baeriswyl
Michel Droz
Andreas Malaspinas
Piero Martinoli

TABLE OF CONTENTS

# THE PROTEIN AS A PHYSICS LABORATORY

Hans Frauenfelder
Department of Physics
University of Illinois at Urbana-Champaign
1110 West Green Street, Urbana, IL 61801

Why should physicists be interested in biomolecules?  One reason is that
physics and in particular physical techniques have had, and still have, a great
impact on biological sciences.  A prime example is X-ray diffraction which in the
hands of Max Perutz and John Kendrew led to the elucidation of the three-
dimensional structure of proteins.  A second reason is the fact that proteins are
beautifully designed laboratories in which many physics problems can be studied.
A few years ago I had dinner with Stan Ulam at the Los Alamos Inn.  After telling
him about our work he said:  "I understand what you are saying.  Ask not what
physics can do for biology, ask what biology can do for physics."  In these notes
I will discuss two areas, complexity and reactions, where experiments on proteins
provide new information.  Both of these areas link biomolecules to physics and
chemistry and both contain many unsolved and challenging problems.

## 1.  PROTEINS

Proteins are the structural elements and the machines of life; they form all
the elements and perform the myriads of tasks that a living system needs.[1]  A
brief description of their construction can be found in ref. 2.  Here only the
most sketchy outline is given.  Proteins are built from twenty different building
blocks, amino acids.  Details of the structure of the amino acids are not
important here.  In constructing a protein, nature covalently links of the order
of 100 to 200 amino acids into a linar "polypeptide" chain.  In the proper solvent
the chain spontaneously folds into the working three-dimensional "tertiary"
structure.  The arrangement of the amino acids in the primary sequence completely
determines the tertiary structure and the function of the protein.

A globular protein typically has a molecular weight of the order of 20,000
dalton, a linear dimension of a few nm, and it consists of a few thousand atoms.
Proteins are therefore complex many-body systems, at the border between classical
and quantum mechanics.  They are also disordered in the sense of a Picasso
painting or Beethoven's Grosse Fuge.  One important aspect is the highly

anisotropic arrangement of the forces. Along the polypeptide chain or backbone, the bonds are covalent and are therefore not broken by thermal fluctuations. The three-dimensionl structure is, however, stabilized by hydrogen bonds and Van der Waals forces. These "weak" forces can be spontaneously broken by thermal fluctuations so that the protein is a very flexible and mobile system.

We will consider a particular class, heme proteins. In these molecules, the folded polypeptide chain or globin contains a small organic molecule, protoheme. Protoheme is a roughly spherical molecule of about 1 nm diameter, with an iron atom at the center. Heme proteins perform a wide variety of tasks, from storage and transport of matter and electricity to the catalysis of reactions. The best known heme protein is hemoglobin, the oxygen carrier. We will be concerned mainly with myoglobin, which stores oxygen in the muscles. Myoglobin (Mb) is built from 153 amino acids, has a molecular weight of about 18,000 dalton, contains about 1200 non-hydrogen atoms, and has dimensions of about $3 \times 4 \times 4$ nm.[3] Fig. 1 shows

Fig. 1  Schematic cross section through myoglobin.

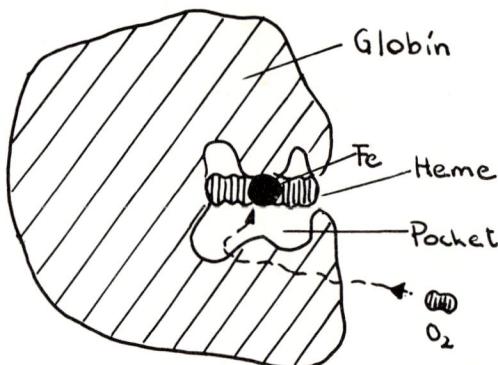

a schematic cross section through Mb. The reversible storage of dioxygen ($O_2$) occurs at the heme iron; we represent it by the equation

$$Mb + O_2 \leftrightarrow MbO_2. \tag{1}$$

This relation appears extremely simple, but it turns out that we know less now than when we started our work about 15 years ago. In fact, the closer one looks at the reactions involved in Eq. (1), the more one appreciates Bohr's favorite Schiller verse:

>"Nur die Fülle führt zur Klarheit,
>Und im Abgrund liegt die Wahrheit."

## 2.  EXPERIMENTAL TECHNIQUES

Biomolecular phenomena are so complex that every available physical and chemical tool must be used for the elucidation of structure and function. We sketch here only two techniques to at least provide some insight into the gathering of the essential experimental data.

2.1 **Flash photolysis.** Flash photolysis is simply a photodissociation experiment. In the standard approach in physics, all one observes is the process of dissociation, as for instance in the photodisintegration of the deuteron. In biological physics, in contrast, both photodissociation and rebinding are observed. Consider for instance carbonmonoxymyoglobin, MbCO, where carbon monoxide is bound to the heme iron of Mb. A laser pulse breaks the bond between the iron atom and the CO molecule. The CO molecule then separates from the iron and at high temperatures from the Mb molecule. Ultimately, however, it will rebind so that the reaction cycle is

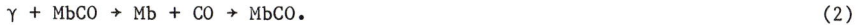

$$\gamma + MbCO \rightarrow Mb + CO \rightarrow MbCO. \tag{2}$$

The reaction can be followd for instance by observation of the optical spectrum near 440 nm, where Mb and MbCO have a very different extinction coefficient (Venous and arterial blood have different color!). It is important to study the reaction (2) over wide ranges in time (fs to Ms), temperature (2–300 K), and pressure (to 2 kbar). The experimental arrangement is simple[3]: A thin MbCO sample is placed into a cryostat with windows. The Fe–CO bond is broken with a short laser flash and the subsequent dissociation and reassociation processes are followed optically. To observe the entire time range, different lasers and approaches are needed.[4-7] Since many protein phenomena are nonexponential in time and cover many orders of magnitude in time, results must nearly always be plotted versus log time. Fig. 2 shows typical rebinding data. The data describe log N(t) versus log t, where N(t) denotes the fraction of Mb molecules that have not rebound a CO molecule at the time t after photodissociation.

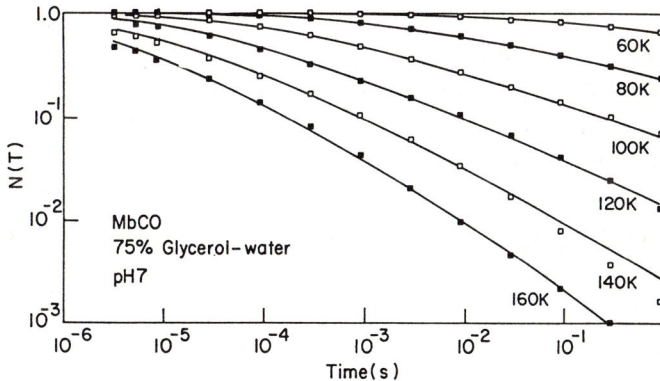

Fig. 2 Time dependence of the binding of carbon monoxide to myoglobin. N(t) is the fraction of Mb molecules that have not rebound a CO at the time t after photodissociation. (The fit is from R.D. Young and S.F. Bowne, J. Chem. Phys. **81**, 3730 (1984)).

2.2 **X-ray diffraction.** As pointed out in the introduction, the determination of the electron density of myoglobin and hemoglobin by Kendrew and Perutz, respectively, was one of the truly fundamental steps in the exploration of biomolecules. A clear and beautiful description of many aspects of the structure determination is given in ref. 8. It turns out, however, that X-ray diffraction is capable of yielding considerably more information than just the average structure. Two applications of particular importance to protein dynamics are the determination of the Debye-Waller factor and of the thermal expansion.

(i) Debye-Waller factor.[9-11] Standard X-ray diffraction yields the average position of each non-hydrogen atom in a protein molecule. If all equivalent atoms sit in exactly the right position, interference is maximal and the Laue spots have maximal intensity. If, however, atoms are spread out or move about their average positions with mean square deviations $\langle x^2 \rangle$, the intensity is reduced by the Debye-Waller factor. From the reduction in intensity of a large number of diffraction spots, the $\langle x^2 \rangle$ for each nonhydrogen atom can be computed. Since these deviations can be caused by the dynamic motion of the protein, dynamic information can be obtained.

(ii) Thermal expansion.[12] Since proteins are highly inhomogeneous and anisotropic systems, a measurement of the thermal expansion as function of position and direction can be expected to provide information about forces and motions. An accurate determination of the coordinates of all atoms as function of temperature provides a very large amount of data concerning the expansion tensor.[12]

## 3. COMPLEXITY

Heme proteins are excellent examples of how proteins can be used as physics laboratories. In fact, even the "simple" myoglobin, Fig. 1, contains at least two different laboratories. One, the entire protein, is well suited for studies of complexity. The other, the heme pocket and the heme group together, permits extensive investigations of reaction theories. In the present section, some of the aspects of complexity will be sketched.

3.1 Nonexponential time dependence. Fig. 2 shows unambiguously that the binding of CO to Mb below about 160 K is nonexponential in time. It can be approximated by a power law,

$$N(t) = N(0) (1 + t/t_o)^{-n}, \tag{3}$$

where $t_o$ and $n$ are temperature-dependent parameters.[3]. Similar nonexponential rebinding occurs in all the heme proteins that have been studied.[13] Such "endless processes" have a long and fascinating history.[14] They were already observed in 1835 by W. Weber in Göttingen[15] and they turn up in a wide variety of fields such as mechanical creep, dielectric relaxation, phosphorescence, luminescence, annealing of radiation damage, NMR, dynamic light scattering, remnant magnetization in spin glasses, and photosynthesis.

Systems exhibiting nonexponential time dependencies have been treated by a wide variety of theoretical approaches, e.g. refs. 14, 16-18. The nonexponential time dependence can be explained by homogeneous or by inhomogeneous processes. Consider a system that consists of a number of subsystems, for instance the individual Mb molecules in a sample. In a homogeneous system, all subsystems are identical and each subsystem exhibits nonexponential time dependence. In an inhomogeneous system, each subsystem can have exponential behavior, but with different rates. The ensemble then shows the nonexponentiality. Remarkably enough, proteins show both types of behavior. The homogeneous case will be discussed somewhat later. In the binding of CO to Mb, we have shown conclusively by repeated photodissociation ("hole burning in time") that the Mb sample must be inhomogeneous.[3,19] Each protein molecule can be characterized by a single rate coefficient. Assume that the rate coefficient k is determined by an Arrhenius relation,

$$k(H) = A \exp(-H/k_B T),   \qquad (4)$$

where H is the height of the barrier governing the reaction. The observed binding process can be fitted by a linear superposition of exponential terms,

$$N(t) = \int dH \, g(H) \exp\{-k(H)/t\},   \qquad (5)$$

where g(H)dH is the probability of having a Mb molecule with barrier height between H and H + dH. Inverting the Laplace transform Eq. (5) (not trivial) with Eq. (4) yields the probability distribution g(H) and values of the preexponential A for each protein-ligand combination. Values of A are typically of the order of $10^9 \, s^{-1}$; g(H) is characteristic for the protein-ligand combination.[3,13]

   3.2  Conformational substates.  Why do different protein molecules with the same primary sequence possess different activation enthalpies H at low temperatures? The simplest explanation is based on the complexity of protein folding and protein structure. Folding is unlikely to lead to a unique tertiary structure. The protein structure is so flexible and so complex that small changes in the structure and the arrangement of the weak bonds and of the water molecules on the outside of the protein are unlikely to change the total binding energy of the protein by much. We therefore assume that a given protein, say sperm whale myoglobin, can exist in a large number of conformational substates (cs).[3,9-11] All conformational substates have the same overall structure, but differ in smaller features. All cs perform the same function, e.g. binding of dioxygen, but may have different rates.

   The concept of conformational substates, introduced in 1973[20], is analogous to the concept of energy valleys in spin glasses.[21] Each substate is a valley in the Gibbs energy surface, separated by high barriers from other valleys. At temperatures below about 180 K, a protein will remain frozen in a particular cs; above about 200 K, a protein will fluctuate from cs to cs. All present experi-

mental evidence is consistent with the concept of substates.  Particularly
striking evidence comes from the Debye-Waller factor.[9-11]

As pointed out above, different substates have different values of the
activation enthalpy H for the binding of CO and $O_2$.  Different substates thus have
different properties and this fact may be analogous to replica-symmetry breaking
in the theory of spin glasses.

3.3  States and substates.  The existence of conformational substates leads
to some new features.  In order to perform a function, a protein must be able to
exist in more than one state.  Myoglobin, for instance, can be in the liganded or
the unliganded state, MbCO and Mb.  Cytochrome c, an electron carrier, can be in
an oxidized and a reduced state.  Since each of these states can assume a
multitude of cs, we must distinguish two different types of motions, equilibrium
fluctuations (EF) and nonequilibrium motions.  EF lead from one substate to
another.  The nonequilibrium motions lead from one state to another.  Since they
are involved in the function of the protein or enzyme, we call them functionally
important motions, or fims.

Equilibrium fluctuations and fims are related by fluctuation-dissipation
theorems.[22-25]  The theorem is, of course, only valid if fluctuations and
dissipative motions cover essentially the same substates.

3.4  Nonergodicity and time scales.[26]  In the application of physical
concepts to proteins and in the extraction of new concepts from biomolecular
experiments, the time scales must be considered.  Assume that a protein can hop
from cs to cs with a rate $k_r = 1/\tau_r$, where $\tau_r$ is the hopping (relaxation) time.
The response of the system to an experimental observation depends on $\tau_r$ and on
the characteristic time $t_{obs}$ of the observation.  If $\tau_r \ll t_{obs}$, the system passes
through all substates during the observation and it appears as ergodic.  If
$\tau_r \gg t_{obs}$, each subsystem is frozen into a particular substate during the
experiment and the system appears as nonergodic.  In general $\tau_r$ is a strong
function of temperature and the properties of the system as seen by a particular
observation will depend on T.

3.5  Proteinquakes.[27]  The shift of tectonic plates in the earth can lead to
stress and the build-up of strain energy.  An earthquake occurs when the stress is
relieved and the strain energy dissipated, resulting in a permanent deformation
and the emission of shear and pressure waves.  In a protein, stress is established
for instance when CO binds to the heme iron in Mb.  When photodissociation breaks
the bond between the iron atom and the CO molecule, the stress is relieved and the
protein changes from the liganded to the unliganded structure.  We call the
rearrangement after the bond breaking a proteinquake.  Progress of the quake can
be followed by monitoring suitable spectroscopic markers.

The proteinquake following the photodissociation of MbCO, monitored by a
number of techniques[27], implies that the release of the strain energy occurs in a

sequence of about four steps. While the details remain to be studied, it is plausible to assume that the quake starts with the motion of the heme iron and the heme and then propagates outward until the entire molecule, including the hydration shell, is rearranged. The first phase of the quake, fim 4, occurs even at 4 K and is extremely fast. The second phase, fim 3, probably takes place near 20 K. The third phase, fim 2, starts near 20 K and extends to at least 120 K. The final phase, fim 1, occurs near 180 K.

Fims 1 and 2 are both nonexponential in time. Fim 2 has so far been investigated in most detail, because it can be monitored by the shift of a small charge-transfer band near 760 nm. The band shifts without noticeably broadening and without exhibiting an isosbestic point. This fact demonstrates that the relaxation must be of the homogeneous type and that a considerable number of intermediate states are involved.

3.6 A hierarchy of substates. The occurrence of several phases in the proteinquake and the nonexponential time dependencies observed in ligand binding and in fims 1 and 2 together lead to a hierarchical model for protein substates. X-ray diffraction demonstrates that the difference in structure between MbCO and Mb is small.[28] The same or similar substates are consequently involved in EF and fims. The existence of four fims then implies four tiers of substates which we denote by $cs^1$ to $cs^4$. The resulting hierarchy of substates is shown in Fig. 3. The energy valley at the top represents one state, for instance Mb. Mb can exist in a large number of conformational substates of the first tier, $cs^1$, separated by high mountains. Each valley in the first tier is divided into $cs^2$, with smaller barriers. The furcation continues, with increasingly smaller barriers.

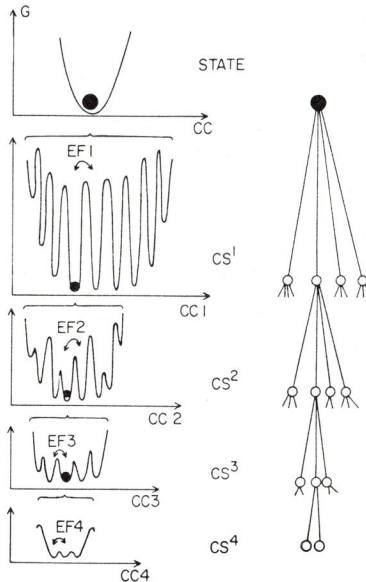

Fig. 3 Hierarchical arrangement of conformational substates in myoglobin. Left: schematic arrangement of energy surfaces. Right: tree diagram. G is the Gibbs energy, cc a conformational coordinate, cs denotes substates. (After ref. 27.)

The hierarchical arrangement of substates leads to a pronounced dependence of the protein motions on temperature. Motions of tier 1 occur only above about 180 K while fluctuations in tier 4 take place even below 1 K. They must consequently involve tunneling.[29,30]

### 3.7 Ultrametricity.

Nonexponential relaxation leads naturally to hierarchical models.[31-33] In 1983 Mézard and collaborators found that the topology of hierarchical models can be ultrametric.[21,34] To briefly discuss ultrametricity we note that only the wells at the bottom of the diagram in Fig. 3 represents reality. A protein will always be in one of the lowest wells; the higher wells only label the system. Consider three points A, B, and C, corresponding to three instantaneous situations of the protein described by Fig. 3. Denote the distance between A and B by AB, where the distance can for instance be the time it takes the system to go from A to B. An ultrametric space is defined by the relation

$$AB \leqslant BC = CA. \tag{6}$$

Triangles can be equilateral or isosceles, with AB smaller than the two equal sides.

We do not yet know if proteins are indeed ultrametric, but the results obtained so far suggest that the possibility exists. If proteins are ultrametric, studies of Brownian motion on ultrametric lattices[33-36] may help elucidate the nonexponential time dependence of protein relaxation.

### 3.8 Proteins, glasses, and spin glasses.

As pointed out by Gérard Toulouse, proteins may be the missing link between glasses and spin glasses. In glasses and proteins, the stochastic variable is the atomic coordinate, while in spin glasses it is the spin. In spin glasses and proteins, the ground state is multiply degenerate, while in glasses the ground state is the crystal.

## 4. REACTIONS

Proteins do nearly all the work in living systems and most of this work involves reactions. The reacions can involve the storage and transport of matter, electricity, or energy or they may occur in the transformation of light energy to chemical energy or chemical energy into motion. A reliable theory of chemical reactions consequently is extremely important for a deeper understanding of protein function. A full theory of chemical reactions, in turn, necessarily involves physics. It is therefore interesting that the protein is an excellent laboratory for the study of reactions. We will describe here only a few aspects of this challenging field. It is likely that more detailed studies of the reaction of and within proteins will provide more incentives for improving reaction theory.

4.1 The Kramers theory. Neither the Arrhenius relation, Eq. (4), nor the well-known Eyring equation[37] contain the viscosity of the medium in which the reaction takes place. Detailed studies of an enzyme reaction[38] and of the binding of CO (Eq. (2))[39] show, however, that the solvent viscosity affects the reaction rate. At first, the viscosity dependence appears to suggest a diffusion-limited reaction. After eliminating this possibility a more important and interesting explanation emerges: Viscosity is proportional to friction. Friction is not an atomic concept, it describes phenomenologically the effect of the exchange of energy and momentum with invisible coordinates. H. A. Kramers introduced friction and fluctuations into reaction theory in 1940.[40] He showed that the rate coefficient is proportional to viscosity at low damping and inversely proportional to viscosity at high damping. The equations of Kramers are better founded, give a better account of reactions and lead to more reasonable values of activation enthalpies and entropies than the Eyring relation. Despite these advantages only a few theoreticians took Kramers' approach seriously and the experimental physicists and chemists neglected it entirely. Within the last few years, the approach of Kramers has become popular (for reviews and references, see 41,42) and its validity has been verified experimentally.[43] It is amusing to note that proteins provided one of the laboratories where the approach of Kramers was resurrected.

4.2 Bond formation at the heme. In Section 3.1 we discussed the binding of CO to Mb without giving any molecular details. Here we return to the problem and show in Fig. 4 the main structural elements involved in the binding at the heme and also a corresponding potential. We denote with B the protein state with the CO in the heme pocket and with A the bound state MbCO. His indicates the distal histidine, the amino acid that links the heme iron to the protein backbone at the F helix. In A, the heme is planar, the iron has spin 0 and is very close to the

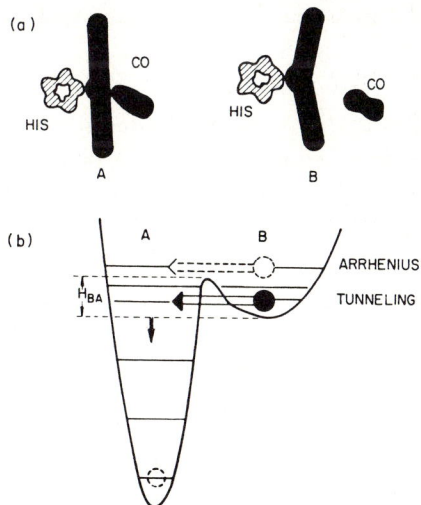

Fig. 4 (a) The two states involved in the binding of CO and $O_2$ to the heme iron in the heme proteins. In A, CO is bound, in B CO is free in the heme pocket. (b) Schematic potential for the covalent binding step B → A.

heme plane. In B, the heme is domed, the iron has spin 0 and lies about 50 pm out of the mean heme plane. The bond formation B → A is represented in the potential diagram as motion of the system from the shallow pocket B to the deep well A. The barrier height H in Eq. (4) is indicated in Fig. 4 and g(H) in Eq. (5) refers to the probability of finding different barrier heights in different protein molecules.

The "laboratory for reaction dynamics" in Fig. 4 yields information on a surprising number of features. We discuss the most important ones of these in the following Sections.

4.3  Tunnel effect. Fig. 4 suggests that the transition B → A should be dominated by quantum-mechanical tunneling below a critical temperature. In the simplest model, the rate coefficient for tunneling of a system with mass M through a parabolic barrier of height H and width d can be written as

$$k_t = A_t \exp\{-\pi d[2MH]^{1/2}/2\hbar\}. \qquad (7)$$

$A_t$ is a preexponential factor. With Eqs. (4) and (7) and setting $A = A_t$, it is easy to estimate that tunneling should become dominant near 20 K.[44] Indeed, the rate for binding becomes essentially temperature independent near 10 K.[45]

Since most tunneling experiments yield only one temperature-independent value of $k_t$, it is impossible to determine the two parameters d and $A_t$ separately. In heme proteins, the distributed barriers described in Section 3.1 permit a separation! The idea is simple. Tunneling and classical Arrhenius motion are but two facets of the same process and they are governed by the same barrier. At temperatures above about 40 K, tunneling can be neglected, the rate coefficient k(H) in Eq. (5) is given by the Arrhenius relation Eq. (4), and g(H) can be determined. Below about 40 K tunneling dominates, g(H) is now known, and the measured N(t) permits a determination of k(H). From k(H), d and $A_t$ can be found.[46]

The second criterion of tunneling is a pronounced isotope effect, characterized by the mass M in Eq. (7). It is experimentally hopeless to observe the tunnel effect with a measurement of the type shown in Fig. 2. Two separate measurements with different isotopes would be required. Because of the nonexponential time dependence of rebinding, the resulting rates would not be accurate enough to characterize an isotope effect cleanly. Rebinding can, however, also be monitored by looking at the stretching frequencies of the bound CO molecules. Since different isotopes have different stretching frequencies, two isotopes can be observed in the same sample and the same measurement. The result is unambiguous; the isotope effect for the pair $^{12}CO$ and $^{13}CO$ is as expected.[47]

4.4  Tunneling problems. Lest it appear that tunneling in heme proteins is well understood, we state here some open problems. (i) While the replacement of $^{12}C$ by $^{13}C$ yields a decrease of the tunneling rate by the expected amount, the replacement of $^{16}O$ by $^{18}O$ actually changes the rate less.[47] This observation demonstrates that CO does not tunnel as a point particle, but that the structure of the tunneling system is important. (ii) Tunneling becomes temperature-independent already at about 20 K. From comparisons with librational tunneling we would expect tunneling to be proportional to temperature down to below 1 K. A possible explanation is again the structure. If the heme group is partially isolated, the phonon spectrum may contain not only an upper, but also a lower Debye cut-off.[46] (iii) In the binding of CO to some proteins, for instance carboxymethylated cytochrome c[48] and horseradish peroxidase[49], we have observed a rebinding process that is very fast even at 4 K, is exponential in time, and appears to proceed by tunneling up to at least 100 K. We have not yet been able to explain the observed temperature dependence in a satisfactory way. (iv) The complete theory of tunneling has to include the coupling to environmental degrees of freedom. While considerable progress in this direction has been made, much work remains to be done.[42,50] This short outline of problems indicates that much experimental and theoretical work on tunneling in proteins remains to be done.

  4.5  Steric and electronic control.[41]  So far we have neglected one major problem: What controls the rate of binding in the step B → A shown in Fig. 4? Two main possibilities exist, steric or electronic control. In the first case, often called adiabatic, control is exerted by steric features. In the second case, usually denoted as non-adiabatic, control is dominated by an electronic matrix element. Steric (adiabatic) reactions are described by the Kramers approach (Section 4.1). Electronic (nonadiabatic) transitions, where the system has to move from one electronic surface to another, are usually treated by an approach due to Landau[51], Zener[52], and Stueckelberg.[53] The probability $P(V_{el})$ of remaining on the adiabatic surface depends on the strength $V_{el}$ of the electronic matrix element and is given by

$$P(V_{el}) = 1 - \exp\{-\pi\gamma_{LZ}/2\}. \tag{8}$$

Here $\gamma_{LZ}$ is the adiabaticity parameter, which is proportional to $V_{el}^2$. For large $V_{el}$, $P = 1$, for small $V_{el}$, P is proportional to $V_{el}^2$. In general, the preexponential factor A in Eq. (4) can now be written as

$$A = \nu \ \kappa(\eta) \ \exp\{S^*/R\} \ P(V_{el}) \tag{9}$$

where $\nu$ is a characteristic frequency of the order of $10^{12} \ s^{-1}$, $\kappa(\eta)$ gives the reduction of the barrier transmission due to friction, $\exp\{S^*/R\}$ is the activation entropy factor, and $P(V_{el})$ characterizes the decrease of the rate due to electronic effects.

  One puzzle posed by the binding of CO and $O_2$ to heme proteins[41] can be discussed in terms of Eq. (9): Both ligands bind with essentially the same rate at low temperatures and for both the preexponential factor A is about $10^9 \ s^{-1}$. Free CO and $O_2$ molecules possess very different electronic properties, however.

CO has spin 0, $O_2$ has spin 1, and the transition B → A requires a spin change 2 → 0 for the iron atom. The transition consequently should be first-order in the spin-orbit interaction for $O_2$, but second order for CO: $O_2$ should bind much faster than CO. Why do the two bind nearly equally fast? A second question is posed by the small value of A. The detailed discussion[41] of the problems leads to a number of conclusions: (i) The factor $P(V_{el})$ can be affected by friction and consequently the approaches of Kramers and of Landau-Zener-Stueckelberg must be combined. It turns out that friction can make a nonadiabatic transition appear adiabatic. The change is, however, not large enough to explain the observed $O_2$-CO equality. (ii) The most likely source of the small value of A is the entropy factor. In binding, the number of states is drastically reduced and this reduction leads to the small A. (iii) The most likely cause for the near equality of $O_2$ and CO is the influence of the protein structure which may produce an intermediate state in the step B → A.

The problems discussed here do not exhaust the results obtained with proteins. They should show, however, that proteins indeed form a physics laboratory in which the most interesting problems appear unexpectedly and where physicists, chemists, and biologists can join forces and can learn from each other.

## ACKNOWLEDGEMENTS

This work was supported by Grant PCM81-09616 from the National Science Foundation and by Grant PHS GM18051 from the Department of Health and Human Services.

## REFERENCES

1.  L. Stryer, Biochemistry, W. H. Freeman and Company, San Francisco, 1981.
2.  H. Frauenfelder, Helv. Phys. Acta 57, 165-187 (1984).
3.  R. H. Austin, K. W. Beeson, L. Eisenstein, and H. Frauenfelder, Biochemistry 14, 5355-5373 (1975).
4.  G. H. Fleming, Ann. Rev. Phys. Chem. 37, 81-104 (1986).
5.  Ultrafast Phenomena IV, D. H. Auston and K. B. Eisenthal, Eds., Springer 1984.
6.  Ultrafast Phenomena V, Springer, 1986.
7.  R. H. Austin, K. W. Beeson, S. S. Chan, P. G. Debrunner, R. Downing, L. Eisenstein, H. Frauenfelder, and T. M. Nordlund, Rev. Sci. Instr. 47, 445-447 (1976).
8.  R. E. Dickerson and I. Geis, Hemoglobin: Structure, Function, Evolution and Pathology, Benjamin/Cummings, 1983.
9.  H. Frauenfelder, G. A. Petsko, and D. Tsernoglou, Nature 280, 558-563 (1979).
10. H. Hartmann, F. Parak, W. Steigemann, G. A. Petsko, D. Ringe Ponzi, and H. Frauenfelder, Proc. Natl. Acad. Sci. USA 79, 4967-4971 (1982).
11. G. A. Petsko and D. Ringe, Ann. Rev. Biophys. Bioeng. 13, 331-371 (1984).

12. H. Frauenfelder, H. Hartmann, M. Karplus, I. D. Kuntz, Jr., J. Kuriyan, F. Parak, G. A. Petsko, D. Ringe, R. F. Tilton, Jr., M. L. Connolly, and N. Max, Biochemistry, in press.
13. F. Stetzkowski, R. Banerjee, M. C. Marden, D. K. Beece, S. F. Bowne, W. Doster, L. Eisenstein, H. Frauenfelder, L. Reinisch, E. Shyamsunder, and C. Jung, J. Biol. Chem. 260, 8803-8809 (1985).
14. J. T. Bendler, J. Stat. Phys. 36, 625-637 (1984).
15. W. Weber, Götting. Gel. Anz. p. 8 (1835), Annalen der Physik und Chemie (Poggendorf) 34, 247 (1835).
16. G. Williams and D. C. Watts, Trans. Farad. Soc. 66, 80 (1970).
17. E. W. Montroll and J. T. Bendler, J. Stat. Phys. 34, 129-162 (1984).
18. J. Klafter and M. E. Shlesinger, Proc. Natl. Acad. Sci. USA 83, 848-851 (1986).
19. H. Frauenfelder, in Structure and Dynamics: Nucleic Acids and Proteins, Adenine Press, 369-376 (1983).
20. R. H. Austin, K. Beeson, L. Eisenstein, H. Frauenfelder, I. C. Gunsalus, and V. P. Marshall, Phys. Rev. Letters 32, 403-405 (1974).
21. M. Mézard, G. Parisi, N. Sourlas, G. Toulouse, and M. Virasoro, Phys. Rev. Letters 52, 1156-1159 (1984).
22. L. Onsager, Phys. Rev. 37, 405-426 (1931).
23. H. B. Callen and T. A. Welton, Phys. Rev. 83, 34-40 (1951).
24. R. Kubo, Rep. Progr. Phys. 29, 255-284 (1966).
25. P. Hänggi, Helv. Phys. Acta 51, 202-219 (1979).
26. R. G. Palmer, Adv. Phys. 31, 669-735 (1982).
27. A. Ansari, J. Berendzen, S. F. Bowne, H. Frauenfelder, I. E. T. Iben, T. B. Sauke, E. Shyamsunder, and R. D. Young, Proc. Natl. Acad. Sci. USA 82, 5000-5004 (1985).
28. S. E. V. Phillips, J. Mol. Biol. 142, 531-554 (1980).
29. V. I. Goldanskii, Yu. F. Krupyanskii, and V. N. Fleurov, Doklady Akad. Nauk SSSR 272, 978-981 (1983).
30. G. P. Singh, H. J. Schink, H. V. Lohneysen, F. Parak, and S. Hunklinger, Z. Phys. B55, 23-26 (1984).
31. M. F. Shlesinger and E. W. Montroll, Proc. Natl. Acad. Sci. USA 81, 1280-1283 (1984).
32. R. G. Palmer, D. L. Stein, E. Abrahams, and P. W. Anderson, Phys. Rev. Lett. 53, 958-961 (1984).
33. B. Huberman and M. Kerszberg, J. Phys. A18, L331-336 (1985).
34. R. Rammal, G. Toulouse, and M. A. Virasoro, Rev. Mod. Phys. 58, 765-788 (1986).
35. A. T. Ogielsky and D. L. Stein, Phys. Rev. Letters 55, 1634-1637 (1985).
36. A. Blumen, J. Klafter, and G. Zumofen, J. Phys. A19, L77-84 (1986).
37. S. Glasstone, K. J. Laidler, and H. Eyring, The Theory of Rate Processes, McGraw-Hill, New York (1941).
38. B. Gavish and M. M. Werber, Biochemistry 18, 1269 (1979).
39. D. Beece, L. Eisenstein, H. Frauenfelder, D. Good, M. C. Marden, L. Reinisch, A. H. Reynolds, L. B. Sorensen, and K. T. Yue, Biochemistry 19, 5147-5157 (1980).
40. H. A. Kramers, Physics 7, 284 (1940).
41. H. Frauenfelder and P. G. Wolynes, Science 229, 337-345 (1985).
42. P. Hänggi, J. Stat. Phys. 42, 105-148 (1986).
43. G. R. Fleming, S. H. Courtney, and M. W. Balk, J. Stat. Phys. 42, 83-104 (1986).
44. V. I. Goldanskii, Dokl. Akad. Nauk SSSR 124, 1261 (1959).
45. N. Alberding, R. H. Austin, K. W. Beeson, S. S. Chan, L. Eisenstein, H. Frauenfelder, and T. M. Nordlund, Science 192, 1002-1004 (1976).
46. H. Frauenfelder, in Tunneling in Biological Systems, Academic Press, 627-649 (1979).

47. J. O. Alben, D. Beece, S. F. Bowne, L. Eisenstein, H. Frauenfelder, D. Good, M. C. Marden, P. P. Moh, L. Reinisch, A. H. Reynolds, and K. T. Yue, Phys. Rev. Letters 44, 1157–1160 (1980).

48. N. Alberding, R. H. Austin, S. S. Chan, L. Eisenstein, H. Frauenfelder, D. Good, K. Kaufmann, M. Marden, T. M. Nordlund, L. Reinisch, A. H. Reynolds, L. B. Sorensen, G. C. Wagner, and K. T. Yue, Biophys. J. 24, 319–334 (1978).

49. W. Doster, S. F. Bowne, H. Frauenfelder, L. Reinisch, and E. Shyamsunder, J. Mole. Biol. in press.

50. A. J. Leggett, S. Chakravarty, A. T. Dorsey, M. P. A. Fisher, A. Garg, and W. Zwerger, Rev. Mod. Phys., January 1987.

51. L. Landau, Sov. Phys. 1, 89 (1932); Z. Phys. Sov. 2, 1932 (1932).

52. C. Zener, Proc. Roy. Soc. Ser. A137, 696 (1932).

53. E. G. C. Stueckelberg, Helv. Phys. Acta 5, 369 (1932).

# THE PHYSICS OF EVOLUTION

Manfred Eigen
Max-Planck-Institut
für Biophysikalische Chemie

Am Fassberg
D-3400 Göttingen, FRG

The Darwinian concept of evolution through natural selection has been
revised and put on a solid physical basis, in a form which applies to
self-replicable macromolecules. Two new concepts are introduced: 'sequence space' and 'quasi-species'. Evolutionary change in the DNA- or
RNA-sequence of a gene can be mapped as a trajectory in a sequence
space of dimension $\nu$ , where $\nu$ corresponds to the number of change-
able positions in the genomic sequence. Emphasis, however, is shifted
from the single surviving wildtype, a single point in the sequence
space, to the complex structure of the mutant distribution that consti-
tutes the quasi-species. Selection is equivalent to an establishment
of the quasi-species in a localized region of sequence space, subject
to threshold conditions for the error rate and sequence length. Arrival
of a new mutant may violate the local threshold condition and thereby
lead to a displacement of the quasi-species into a different region of
sequence space. This transformation is similar to a phase transition;
the dynamical equations that describe the quasi-species have been shown
to be analogous to those of the two-dimensional Ising model of ferro-
magnetism. The occurrence of a selectively advantageous mutant is biased
by the particulars of the quasi-species distribution, whose mutants are
populated according to their fitness relative to that of the wildtype.
Inasmuch as fitness regions are connected (like mountain ridges) the
evolutionary trajectory is guided to regions of optimal fitness. Evolution experiments in test tubes confirm this modification of the simple
'chance and law' nature of the Darwinian concept. The results of the
theory can also be applied to the construction of a machine that provides optimal conditions for a rapid evolution of functionally active
macromolecules.

An introduction to the physics of molecular evolution by the author
has appeared recently [1]. Detailed studies of the kinetics and mechanisms of replication of RNA, the most likely candidate for early evolution [2,3], and of the implications on natural selection have been giv-
en in [4,5]. The quasi-species model has been constructed in [6,7] us-

ing the concept of sequence space. Subsequently various methods have been invented to elucidate this concept and to relate it to the theory of critical phenomana [8-19]. The instability of the quasi-species at the error threshold is discussed in [20]. Evolution experiments with RNA strands in test tubes are described in [21,22].

## References

1. Eigen, M., Chemica Scripta 26B, 13 (1986).
2. Eigen, M., and Winkler-Oswatitsch, R., Naturwissenschaften 68, 217 (1981).
3. Eigen, M. and Winkler-Oswatitsch, R., Naturwissenschaften 68, 282 (1981).
4. Biebricher, C.K., Eigen, M. and Gardiner, W.C., Biochemistry 22, 2544 (1983).
5. Biebricher, C.K., Eigen, M. and Gardiner, W.C., Jr., Biochemistry 23, 3186 (1984); 24 (1985).
6. Eigen, M., Naturwissenschaften 58, 465 (1971).
7. Eigen, M., and Schuster, P., Naturwissenschaften 64, 541 (1977); 65,7 (1978); 65, 341 (1978).
8. Eigen, M., Adv. Chem. Phys. 33, 211 (1978).
9. Thompson, C.J. and McBride, J.L., Math.Biosci. 21, 127 (1974).
10. Jones, B.L., Enns, R.H. and Rangnekar, S.S., Bull.Math.Biol. 38, 15 (1976).
11. Jones, B.L., J.Math.Biol. 6, 169 (1978).
12. Schuster, P. and Sigmund, K., Ber.Bunsenges,Phys.Chem. 89, 668 (1985).
13. Swetina, J. and Schuster, P., Biophys.Chem. 16, 329 (1982).
14. Hofbauer, J. and Sigmund, K., Evolutionstheorie und dynamische Systeme, Paul Parey, Berlin and Hamburg (1984).
15. Feistel, R. and Ebeling,W., Bio Systems 15, 291 (1982); and Ebeling, W., Engel, A., Esser, B. and Feistel, R., J.Statist.Phys. 37, 314, 369 (1984).
16. McCaskill, J.S., J.Chem.Phys. 80(10), 5194 (1984).
17. McCaskill, J.S., Biol.Cybernet. 50, 63 (1984).
18. Rumschitzki, D., J.Chem.Phys. (in the press).
19. Leuthäusser, I., J.Chem.Phys. 84, 1884 (1986).
20. Eigen, M., Ber.Bunsenges.Phys.Chem. 89, 658 (1985).
21. Sumper, M. and Luce, R., Proc.Natl.Acad.Sci.USA 72, 162 (1975).
22. Biebricher, C.K., Eigen, M. and Luce, R., J.Mol.Biol. 148, 369 (1981); 148, 391 (1981).

# SYMMETRIES IN BIOLOGY

Yves BOULIGAND

E.P.H.E. & C.N.R.S., 67, rue Maurice-Gunsbourg,
94200 Ivry-sur-Seine (F.).

SUMMARY

This topic being extremely large, this presentation is only a key to literature, with some indications on recent trends in the study of symmetries and symmetry breakings in biological morphogenesis. Symmetry problems are essential in condensed matter physics and, for instance, in research on solid and liquid crystals. Living matter can be considered as a mosaic of solids, of liquids and of a large series of intermediate states, which often are liquid crystals or close analogues of liquid crystals.

Is it possible to develop symmetry studies on biological systems, as do physicists in their own field ? This question was given a positive answer at the molecular level by Louis Pasteur in the nineteenth century and all further studies have confirmed this pioneer work.

The problems considered here concern higher levels of organization and morphogenesis of structures elaborated by considerable sets of cells. For instance, the shapes of organs and of individuals are elaborated mainly by the production of fibrous networks made of various biopolymers. Most classical examples of these networks are found in the integument, in the connective tissue and in the skeletal system. Morphogenesis of such networks results from the activity of cells secreting polymers and from a self-assembly mechanism, resembling a transition from an isotropic state to a liquid crystal in concentrated solutions of these polymers. These ordered secretions are stabilized either by chemical cross-linking between polymers or by microcrystals forming within the liquid crystalline phase. This gives solid or supple systems, showing in their organization most structures and symmetries of liquid crystals.

Liquid crystals contain several types of singular points and lines, whose distribution is often regular and this leads to the differentiation of characteristic textures and shapes. Such architectures also exist in the biological counterpart of liquid crystals. Chiral components and helical polymers are essential in the formation of highly elaborated morphologies of liquid crystals and this is probably one reason why enantiomers rather than racemates or non active components are adopted in living systems.

INTRODUCTION

The subject of symmetry really begins in biology with the contribution of Louis Pasteur in the second half of the nineteenth century [1]. Pasteur showed that optically active isomers are characteristic components of living beings and he underlined the difficulty of asymmetrical synthesis, a problem related to the origin of life. Literature dealing with this question is now extensive, whereas another important topic is much less studied and concerns the reasons why the involvement of chiral molecules is an essential prerequisite of life processes. Recent research on liquid crystals and their biological analogues provides remarkable illustrations of the architectural role of chiral molecules in most complex morphogeneses.

It is also in the second half of the nineteenth century, that began studies about symmetries of whole organisms and particularly with works of Ernst Haeckel [2]. Symmetry problems are encountered at all levels of biological organization, from molecules to cells and from cells to highly organized multicellular organisms.

The birth of life resembles a symmetry breaking. Developmental biology affords examples of differentiations which break certain symmetries in embryos and there are also changes, in the course of evolution, which affect symmetries characterizing certain phyletic groups. Histological and cytological observations demonstrate that chiral liquid crystals or stabilized systems with similar symmetries are widespread in cells and tissues. It appears therefore that symmetry studies at high levels of organization in living systems require the examination of these particular types of order.

SYMMETRY PRINCIPLE AND SPONTANEOUS GENERATION

**Symmetries of causes and effects**

In the last century, Neumann and Curie developed independently important considerations on symmetries, from their studies on crystals. Neumann's principle states that the symmetry elements of any physical property of a medium include those of the medium itself [see 3]. More generally, Curie's principle considers symmetries of causes and effects in physical phenomena and states that dissymmetries of effects are present in causes, whereas effects can be more symmetrical than causes [4].

The symmetry priciples due to Neumann and Curie are fundamental in biology. Before Pasteur demonstrated 'the impossibility of spontaneous generation', in the present biosphere, for microorganisms similar to those observed today, he had studied questions of pure physical chemistry and particularly cristallography of optical isomers of sodium and ammonium salts of tartaric acid. Pasteur's ideas on symmetry preceded those of Curie in several respects and the Curie principle can be stated in terms very close to the thought of Pasteur: <u>there are no spontaneous generations of dissymmetries</u>'. Pasteur and Curie were well aware of the existence of symmetry breakings, but these were supposed to result from very small dissymmetries already present in the causes and able to break an unstable or weakly stable equilibrium. In such situations, very small dissymmetries are strongly amplified.

### Asymmetrical synthesis and the origin of life

A consequence of the Curie principle is the extreme difficulty for chemists to realize asymmetrical syntheses. In the mind of Pasteur, this problem and that of the origin of life were linked. He did not reject the idea of a spontaneous generation of life, favoured by strongly chiral conditions in the prebiotic environment or arising from a considerably amplified symmetry breaking. He observed himself the growth of right and left crystals in equal proportions from racemic solutions of salts of tartaric acid and, despite the symmetry breaking, Curie's principle is verified statistically.

It also happens that racemic solutions of certain compounds cristallize and transform into a single active crystal as shown by Havinga [5]. Upon slow crystallization, the active crystal grows, whereas the mother liquor remains virtually racemic, through a rapid racemization process. Even in such extreme cases, the Curie principle is not really violated, since repeated experiments give both orientations in proportions which do not differ significantly. However, symmetry is broken by the very particular situation of a unique experiment.

Giant crystals of quartz also exist in nature and are right or left, in equal proportions. However certain studies indicate a difference of about 1%, which has to be verified [see 6].

Local symmetry breakings and possibly fundamental dissymmetries of the universe were involved in the selection of the first active compounds useful in elementary forms of life. The development of organisms at the earth surface functions as an amplifier of this primordial dissymmetry and, in a certain sense, life also is a unique experiment.

## SYMMETRY ELEMENTS IN BIOLOGICAL SYSTEMS

### Symmetries at different levels of organization

Many books and reviews afford excellent informations on symmetries of organisms and a remarkable set of illustrations is due to Haeckel [2]. A selection of these pictures is reproduced in 'Growth and Form' by d'Arcy Thompson [7]. Examples of mirror symmetries and of discrete axial symmetries are discussed. Beautiful polyhedral symmetries also exist in viruses [8], in radiolarians [2,7] and in pollens [9].

At the molecular level, most biological components are chiral: oses, aminoacids, phospholipids, etc. Main biopolymers form right-handed helices. Self-assembly of globular proteins leads to structures which are generally helicoidal [rev. in 10]. As will be seen below, there are also examples of helicoidal organization in liquid crystals [11,12] and in their biological analogues [bibliography in 13-15]. Up to this level, there are no mutations leading to an inversion of chirality. . On the contrary, at higher organization levels, when chiral structures depend on relative positions of cells, there is a genetic control and the orientation of gastropod shell is a well known example [16].

Many structures in cells and tissues correspond to ordered media, characterized by symmetry groups, which are those of solid or liquid crystals. Certain tissues contain minerals such as apatite or calcite, which are crystals with definite symmetry groups. There are many other examples of organic or mineral crystals in cells and in the extracellular space of certain tissues. The great majority of ordered structures in living organisms present the characteristic symmetries of liquid crystals, also called mesomorphic states [12-15]. Liquid crystals are anisotropic fluids and one essential example is that of cell membranes, which are well known to be fluid [17-18], with an obvious anisotropy due to the orientation of phospholipids and other important molecules. There are examples of stacked cell membranes, which are strongly reminiscent of smectic phases, namely in rods and cones of retina, in nerve myelin and in white matter of the brain.

**Symmetries of chromatin**

DNA and other double helical nucleic acids give liquid cristalline phases in concentrated solution in water [19-23]. This type of liquid crystal is called 'cholesteric', since it was discovered in numerous cholesterol derivatives. However, similar cholesteric liquids are obtained with various polypeptides, polysaccharides and other chiral polymers. In chromosomes of procaryotes (bacterias and protozoa such as dinoflagellates), the DNA is almost pure and not associated to basic proteins and, in these conditions, it forms cholesteric phases.

The local structure of cholesteric phases is recalled in fig.1. Let us consider a set of equidistant and parallel planes and parallel straight lines in each plane. The orientation of lines rotates by a constant angle from plane to plane, so that the system is regularly twisted. Many biological examples show this twisted stacking of layers and there are for instance cylindrical viruses which form such twisted stratifications. However in general, there are no discrete steps of rotation and the twist is continuous.

Fig. 1. Distribution of molecular orientations in a cholesteric polymer. The The mean direction of polymers rotates by a constant angle from plane to plane. In oblique section, this cholesteric structure draws series of parallel nested arcs. The strucure is continuous in general, and the equidistant planes are used simply to facilitate the drawing.

A local system of Cartesian co-ordinates Oxyz can be introduced, so that the components of a unit vector **n** representing the local orientation of molecules reads:

$$n_x = \cos 2\pi z/p \; ; \; n_y = \sin 2\pi z/p \; ; \; n_z = 0 .$$

The constant p is the helicoidal pitch. The real periodicity is p/2 because **n** and −**n** can be considered as equivalent, since molecules align whatever they are parallel or antiparallel.

In the case of DNA, the double helix diameter is 20 Å and the half helicoidal pitch varies from 800 to 10,000 Å, depending on DNA concentration in water and on the presence of salts. In chromosomes, this half pitch varies only from 800 to 1600 Å. Each chromosome is made of a unique DNA molecule and possibly two before cell division. The length of this chromosome DNA molecule lies between one or several tens of microns and thus the DNA molecule is folded back on itself in numerous points to form a cholesteric rodlet [14].

These elongated chromosomes are suspended within a medium called nucleoplasm, which presents an isotropic symmetry. This situation is schematized in fig. 2. In certain points at the interface separating the two phases, isotropic and cholesteric, the DNA molecule crosses the interface, leaves the cholesteric rodlet and forms large loops in the surrounding isotropic medium. It appears therefore that DNA is present in the two phases, but at very different concentrations.

A micrograph of a thin section of a procaryotic chromosome is shown in Fig.3. The superimposed series of nested arcs indicated in the model of Fig. 1 are well recognizable here. There are no preferential orientations of granular and fibrillar materials present in the surrounding medium.

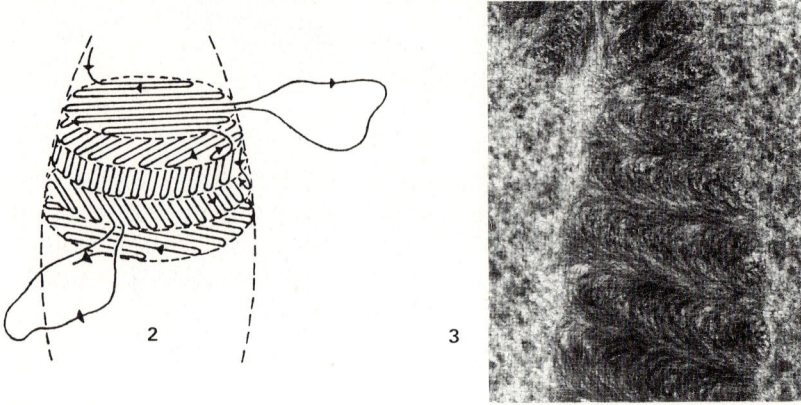

Fig. 2. Hypothetical path of the DNA double helix in a procaryotic chromosome. The succesive planes of backfolding of the polymer are not differentiated in reality, as in Fig.1. Some loops extend in the surrounding isotropic medium or nucleoplasm. Fig. 3. Thin section of a chromosome of Prorocentrum micans, a procaryotic species of the group of Dinoflagellates. The width of the chromosome is about 1 μm.

The chromosomes of procaryotes resemble germs of cholesteric phase in equilibrium with the isotropic phase. Such elongated cholesteric germs were prepared in vitro from purified and sonicated calf thymus DNA [22]. This is an example of liquid crystalline self-assembly, resembling the condensation of chromosomes. The only differences are that DNA molecules used in these experiments are much smaller than the unique chromosome molecule and that proteins associated to chromatin are absent. Condensation of chromosomes or their dilution correspond to the displacement of an equilibrium between cholesteric and isotropic phases. One possible factor of the elongated shape of chromosomes is the anisotropy of the surface tension. Cholesteric structures also exist in the DNA organization of sperm heads [24]. Chromosomes of eucaryotes are very different from those of procaryotes, but many facts suggest a similar liquid crystalline behaviour.

**Symmetries of muscles and skeleton**

In muscles, actin and myosin filaments result from the self-assembly of numerous different proteins; these filaments assemble themselves into myofibrils, the contractile system of muscle cells and myofibrils present the symmetry groups encountered in nematic and in various smectic liquid crystals [14,15]. The liquid character is abolished, since actin and myosin filaments are linked in several ways, but the system is closely related, structurally and physically, to liquid crystals [25-26].

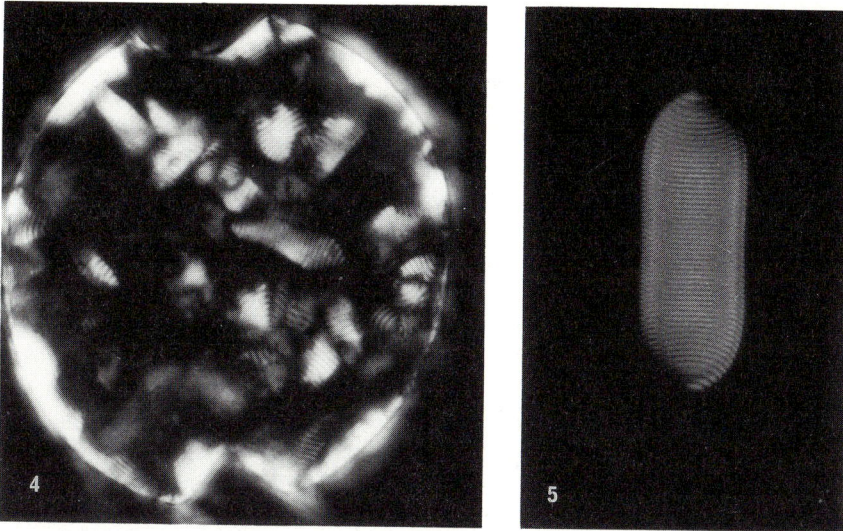

Fig. 4. Small drop of sonicated calf-thymus DNA, in an aqueous solution added with KCl (0.4 M), and concentrated by extremely slow evaporation. The observation between crossed polars shows the formation of elongated cholesteric rodlets (Courtesy of Dr. Livolant [22]). Fig. 5. A similar cholesteric rodlet observed at the isotropic transition in a mixture of cholesterol benzoate and methoxybenzylidene-butylaniline (MBBA), a classical molecule forming liquid crystalline phases at ordinary temperatures.

In connective tissues and in skeletal tissues, triple helical molecules of collagen are assembled into fibrils stabilized by certain cross-links, observed at definite points along fibrils. However, between these cross-links, molecules behave transversally as in a two-dimensional liquid [27] and X-ray diffraction studies confirm this type of order [28]. Collagen fibrils are comparable to certain smectic liquids, but they are only analogues, since they are stabilized and are not really liquid. Collagen form also fibrous systems which are analogues of cholesteric liquid crystals [29]. Several other biopolymers form similar cholesteric analogues of liquid crystals, and a considerable series of derived structures, that we call 'biological plywoods', since their structure is closely related to that of thin laminated wood, with two orthogonal orientations of wood fibers, which alternate.

ROLE OF CHIRAL LIQUID CRYSTALS IN BIOLOGICAL MORPHOGENESIS

**Defects and textures**

A remarkable property of liquid crystals is their ability to arrange spontaneously according to very elaborated morphologies [30]. As in other ordered media, there are singular points and lines in liquid crystals; walls are rare. Owing to the liquid character, these 'defects' adopt regularly distributed positions minimizing the elastic energy. Many of the corresponding textures are easily reproducible, by control of temperature, of pressure, of boundary conditions, particularly angles of molecular anchoring and by applying certain external fields: electric and magnetic fields, temperature gradient etc. [11,12].

In biological analogues of liquid crystals, the characteristic singularities and textures are observed, but there can be slight differences with defects of true liquid crystals. The relationship is extremely narrow between liquid crystals and their biological analogues [31]. Cholesteric liquids and their non liquid analogues in biological systems provide the greatest collection of textures that we know. It is obvious that there are no possible measurements to compare the amplitude of texture variations and our judgement is somewhat subjective. It seems however that chiral molecules provide in a unique manner a considerable repertory of morphologies, particularly with the liquid crystalline phases in which they can be involved.

**Twist and frustration**

Another interesting type of liquid crystal produced by certain chiral molecules corresponds to the 'blue phases' which appear in a narrow range of temperatures between cholesteric and isotropic phases. These blue phases are cubic, and present very particular symmetries [32]. A possible model of one blue phase is represented in Fig. 6. The structure combines three sets of cylinders oriented with their axes parallel to three orthogonal directions, these cylinders being mutually tangent. Molecules are aligned along the cylinder axes and rotate continuously about the radial direction as one moves out, making an angle of 45° on the surface (Fig. 6). This kind of twist can be called 'cylindrical' and differs from the planar twist described in Fig. 1.

The planar twist can be considered as 'frustrated', because twist occurs only along normals to parallel planes in Fig.1 and not within these planes. On the contrary, in the cylindrical situation of Fig. 6, the twist is not frustrated, because it is realized radially, around each molecule, along the cylinder axis and in its vicinity. The energy is locally low within each cylinder, since frustration is releaved. The spaces left between cylinders must be filled with the liquid crystal and this introduces a regular lattice of defects. However, near the isotropic transition, the energy cost of defects is supposed to be very low. This ceases to be true at lower temperatures and the frustrated cholesteric structure replaces the blue phase. The advantage is then that the whole space can be filled with the planar twisted structure, with a much lower density of defects or even no defects.

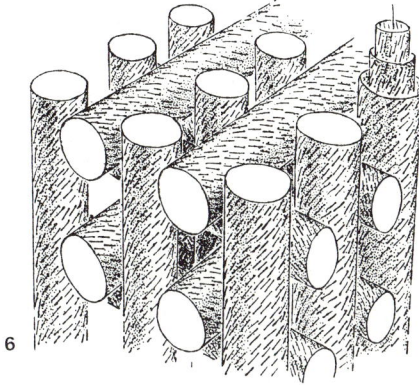

Fig. 6. One of the possible models of blue phase [32]. The twist is not planar as in Fig. 1 but cylindrical, as represented in the cylinder on the right. Small segments at the surface of cylinders correspond to the elongated molecules.

There are examples of biological systems which are closely related to the structure of blue phases. DNA prepared in vitro, in concentrated aqueous solution, can transform slowly from the isotropic state to the cholesteric phase, and there can be intermediate steps, which are liquid crystalline and present remarkable square patterns, reminiscent of blue phases [33]. Another instance is the outer layer, called cuticle, of the skin of certain marine worms [34]. This cuticle is made of two orthogonal sets of coiled cylinders which are twisted bundles of collagen molecules, as represented in Fig. 7. Instead of three orthogonal directions of cylinders, there are only two, and the twisted cylinders of the third direction are replaced by microvilli which are cylindrical cytoplasmic expansions, projecting from epidermal cells and penetrating the orthogonal lattice of collagen, comparable in its principle to a 'plywood'.

## Fibrous networks

There is a considerable variety of 'biological plywoods' which differ by their biochemical composition and by their precise architecture. The more or less solid analogues of cholesteric liquid crystals can be called 'continuously twisted plywoods'. They are very frequent in carapaces of insects and of other arthropods [13]. The twisted layers lie parallel to the epidermis, which secretes them. They are made of chitin (polyacetylglucosamine) associated with proteins and they form extremely resistant exoskeletons. Similar twisted plywoods exist in vertebrate bones, in which collagen fibrils form the organic matrix. Cellulose fibrils also form plywoodlike structures in plant cell walls [13].

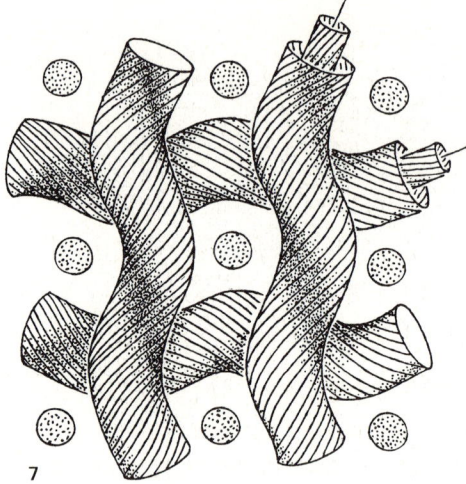

Fig.7 Schematic drawing of two superimposed layers of collagen fibrils in the skin of Paralvinella grasslei. This is a top view; the coiled bundles are parallel to the skin and left-handed, as the twist within the bundles, wheras collagen molecules, which are extremely long polymers, follow right-handed helices. Microvilli are shown penetrating between fibrils (dotted circles).

There are many examples of collagen plywoods in the skin of invertebrates and vertebrates. For instance, in certain sea-cucumbers (holothurians, echinoderms), the dermis is a collagen lattice showing the structure of a stabilized cholesteric phase [35]. In vertebrates, in embryoes and adults, orthogonal networks of collagen are frequent and recognizable very early in the development [36]. In amphibians and reptiles, keratin fibrils form more or less orthogonal layers in surface cells of the epidermis [37].

What we know at the present time about these biological plywoods is that they all result from the transformation of an initial structure, which is a chiral liquid crystal or a very close analogue and that chiral molecules are involved in their morphogenesis [38]. This can be shown from examples, ontogenetic or phylogenetic, showing the transformation of twisted plywoods into orthogonal networks and such changes are frequent.

Muscle cells surrounding the stomach and the intestine are elongated and oriented within superimposed layers, the organization being reminiscent of a plywood structure [39]. Ventricle muscles in the heart of man show a more or less twisted plywood architecture [40]. In the development of such tissues, connective cells secrete, in the first steps of organogenesis, a layered network with different orientations of collagen fibrils and this serves as a matrix for differentiation of myoblasts, which transform into elongated muscle cells and align along collagen fibrils.

## DUALITY OF MORPHOGENESIS

### Self-assembly

Self-assembly studies are among the most fascinating topics of molecular biology. Many cell organelles and viruses can be dissociated into their molecules, which are separated and purified (Reviews in [41,42]). It is then possible to find the physical and chemical conditions, which are appropriate to reassociate them into complete organelles and viruses, with their initial properties and particularly the infectious power of viruses. Bacterial flagella, actin and myosin filaments, microtubules, ribosomes, nucleosomes are the most studied organelles, whose growth corresponds to a self-assembly mechanism. Macromolecules are linked by weak bonds in general in self-assembly and, therefore, enzymes are not always involved in these mechanisms. Each macromolecular subunit occupies a definite position in the whole virus or organelle, exactly as an ion, an atom or a molecule, when it belongs to a crystal lattice.

Self-assembly in biology is a process reminiscent of molecular crystallization. Our presentation shows that several mechanisms of self-assembly are also related to the growth of liquid crystals: cell membranes, procaryotic chromosomes, construction of connective or skeletal structures and also ordering of actin and myosin filaments into myofibrils. In such assembly processes, the subunits are often much larger than usual macromolecules and in muscles, for instance, filaments are aggregates of hundreds of macromolecules. There are also in the cytoplasm of certain living cells 'crystals or liquid crystals of viruses'. This means that the whole viruses are the subunits of a condensed phase which shows the properties of a crystal or of a liquid crystal.

In liquid crystalline self-assembly, subunits do not occupy definite positions as they do in true crystals or in classical self-assembly. The order is mainly orientational. These systems remain fluid in membranes and in chromosomes in general, whereas they are stabilized in muscles and in connective or skeletal structures. The absence of two- or three-dimensional lattices gives a suppleness which is adapted to the elaboration of shapes.

### Role of cells

Morphogenesis of individuals is not limited to self-assembly processes. The essential living unit is the cell. Viruses do not reproduce and do not complete their cycle without a parasitic phase within a host-cell [8]. Certain cell organelles are self-assembled, but cells themselves come from divisions of mother cells, an original process, based on a series of hydrodynamical events observed in protoplam, involving an extremely complicated biochemical and biophysical machinery. In non dividing cells, cytoplasmic streams are also observed and are more or less convective. These currents may have a rôle in the shape determination of cells. Cytoplasmic streams and adhesion of membranes allow many cell types to move. In multicellular organisms, the first steps of morphogenesis are based on cell divisions, on their deformations and on their migrations [36]. All these important phenomena differ from self-assembly, which is limited to morphogenesis of extracellular matrices, of cell membranes and of certain cell organelles, but many of these structures still wait to be reassembled in vitro from their molecular components. Morphogenesis is a dual process involving cell activities and self-assembly.

Studies on symmetries in biology draw their inspiration from crystallography and, more generally, from our knowledge of ordered media, when self-assembled structures are considered. The situation is less simple, when developed structures depend on cell activities [10]. For instance, we indicated above that there are no mutations changing chirality at the molecular level or in the self-assembled structures. On the contrary, the handedness of certain organs depends on the relative positions of cells in the first steps embryogenesis and can be controlled genetically, but the mechanisms are still unknown.

Certain mirror symmetries exist between true twins, developed separately from the first two cells of the egg cleavage. There are also fundamental asymmetries in the human brain etc. At this date, it seems that these chiral macroscopic characters do not depend on presence or absence of certain asymmetrical molecules.

Symmetry breaking in the course of development is reviewed in [10] and [43]. The embryogenesis shows a series of symmetry breakings, which affect either the whole organization or only the self-assembled ordered media in cells and tissues. The study of broken symmetries underlines the duality of morphogenesis: self-assembly and cellular activities. Symmetry breaking in self-assembled systems resmbles what we know in physical systems. There are defects corresponding to locally broken symmetries and examples are numerous in biological ordered media [44]. Certain symmetries are broken globally at certain stages of differentiation of biological ordered media and this resembles to phase transitions. On the contrary, morphogenetic mechanims based on cell activities are still very far from being understood.

Changes in symmetry also occur in the course of evolution, and since there are links between phylogeny and ontogeny, this question is essential. However, our knowledge of this field is limited and purely descriptive. For instance, as indicated above, regular polyhedral shapes exist in radiolarians or in flower pollens, in closely related species and, apparently, there are no possible intermediate steps. This suggests strongly discontinuous evolutions, and sudden changes in the symmetries of whole organisms.

Most penetrant works on symmetries in evolutionary biology are due to Granjean, who mainly studied arthropods, whose bilateral symmetry, internal and external is perfect in general [45]. However, certain minute but well identified organs are asymmetrically present or absent. From statistical studies of these asymmetries, observed at different stages of development and in closely related species, certain general laws can be established and there are clear links with recent results of developmental genetics (Review in [46]).

CONCLUSIONS

Let us come back to our introductory question: What are the reasons making chiral molecules indispensable to life ? Curie gave a crucial indication, when he showed that dissymmetries create phenomena [4]. A remarkable progress illustrating this point was the discovery of the helical growth of crystals. Screw dislocations are indispensable to the growth of crystals, and whatever their handedness. They create at the crystal surface a 'cozy corner' where new sub-units can be added. These locally broken symmetries facilitate a phase transition, which also corresponds to a global symmetry breaking. Screw dislocations are observed in the growth of many mineralized tissues. The apatite crystals of enamel in vertebrates or the nacre of mollusc shells are well known examples. Screw dislocations also are frequent in liquid crystals and in their biological counterpart.

Polymers growing from chiral monomers are helicoidal, in general, and the handedness is well defined. These polymers form cholesteric liquid crystals, whose helicoidal pitches are large relative to the molecular dimensions. From the twist frustration, new orders appear which are reminiscent of the structure of 'plywoods', a frequent organization in the integument, in connective tissues and in the skeleton, the three main systems involved in the elaboration of shapes and their maintenance.

LITERATURE

1. L. Pasteur, Oeuvres Complètes, 1922, Masson, Paris.
2. E. Haeckel, Report on the Scientific Results of the Voyage of H. M. S. Challenger, 18, H. S. M. O., 1887, Leipzig and Wien, 1899; Kunstformen der Natur, Verlag des Bibliographischen Institüt, Leipzig and Wien, 1899.
3. J. F. Nye, Physical Properties of Crystals, 1957, Clarendon Pr., Oxford.
4. P. Curie, Oeuvres Complètes, 1908, Paris, Gauthier-Villars et 1984, Paris, Edition des Archives Contemporaines.
5. E. Havinga, Biochim., Biophys. Acta, 1954, 13, 11.

6. A. and H. Amariglio, X. Duval, Ann. Chim., 1968, 3, 5.

7. D'Arcy W. Yhompson, On Growth and Form, 1917, Cambridge, The Univ. Pr.

8. A.J. Dalton and F. Haguenau ed., Ultrastructure of Animal Viruses and Bacteriophages, an Atlas. 1973, Acad. Pr., N.-Y. and London.

9. H. Mohl, Ann. Sci. Nat., 1835, Paris, sér.II, 3, 148-188; J. Muller, Ann. Missouri Bot. Gard., 66, 593-632.

10. Y. Bouligand, in Symmetries and Broken Symmetries in Condensed Matter Physics, 1981, Boccara N. ed., IDSET, Paris.

11. P.-G. de Gennes, The Physics of Liquid Crystals, 1974, Clar. Pr. Oxford.

12. H. Kelker and R. Hatz, Handbook of Liquid Crystals,1980, Verlag Chemie, Weinheim.

13. Y. Bouligand, Tissue & Cell, 1972, 4, 189-217.

14. Y. Bouligand, Solid States Physics, 1978, Suppl. 14, 259-294.

15. Y. Bouligand, in Liquid Crystalline Order in Polymers,1978, A. Blumstein ed., Acad. Pr.N.-Y., London., 261-297.

16. G.A. Uebbels, J.J. Bezem and C.P. Raven, J. Embryol. Exp. Morph.,1969, 21, 445.

17. D. Chapman, Ann. New-York Acad. Sciences, 1966, 137, 745.

18. S.J. Singer and G. Nicolson, Science, 1972, 175, 720.

19. Robinson C., Tetrahedron,1961, 13, 219.

20. Y. Bouligand, M.-O. Soyer and S. Puiseux-Dao., Chromosoma, 1968, 24, 251.

21. L.S. Lerman, Cold Spring Harbor Symp. Quant. Biol., 1973, 38, 59.

22. F. Livolant, Eur. J. Cell Biol.,1984, 33, 300-311.

23. R.L. Rill, Proc. Nat. Acad. Sci. (U.S.A.), 1986, 83, 342.

24. F. Livolant, Tissue & Cell, 1984, 16, 535-555.

25. E.W. April, P. W. Brandt and G.F. Elliott, J. Cell Biol., 1971, 51, 72-82; 1972, 53, 53-65.

26. G.F. Elliot and E. Rome, Mol. Crystals and Liq. Crystals, 1969, 8, 215-219.

27. D.A. Torchia and D.L. van der Hart, J. Mol. Biol., 1976, 104, 315-321.

28. J. Woodhead-Galloway, D.W.L. Hukins, D.P. Knight, P.A. Machin and J.B. Weiss, J. Mol. Biol., 1878, 118, 567-578.

29. Y. Bouligand, J.-P. Denèfle, J.-P. Lechaire and M. Maillard. Biol. Cell, 1985, 54, 143-162.

30. Y Bouligand, J. Physique, 1972-1974, 33,525-547, 715-736; 34, 603-614,1011-1020; 35, 215-235, 959-981.

31. Y. Bouligand, J. Physique,1975, 36, C1, 331-336.

32. S. Meiboom, J.P. Sethna, P.W. Anderson and W.F. Brinkman, Phys. Rev. Lett., 1981, 46,1216-1219.

33. F. Livolant, J. Physique,1987,in the press.

34. L. Le Pescheux, private communication, manuscript in preparation.

35. Y. Bouligand, in Mesomorphic Order in Polymers, A. Blumstein ed., ACS Symp. Ser., 76, 237-247.

36. L.E.R. Picken, The Organization of Cells and other Organisms, 1961, Clarendon Pr., Oxford.

37. Y. Le Quang-Trong and Y. Bouligand, Bull. Soc. Zool. Fr.,1976,101, 637-645.

38. Y. Bouligand and M.-M. Giraud-Guille,1985, in Biology of Invertebrate and Lower Invertebrate Collagens,1985, A. Bairati and R. Garrone eds., Plenum.

39. W. Bloom and Don W. Fawcett, A Textbook of Histology, 10th edition, 1975, W.B. Saunders.

40. R. Olson, in Handbook of Physiology, Circulation, 1, Chapt. 10, 199-235.

41. A. Lehninger, in Biochemistry. The Molecular Basis of Cell Structure and Function, 2nd ed., 1975, Worth Publ. N.-Y.

42. P. Favard and Y. Bouligand, in 'La Morphogenèse, de la Biologie aux Mathématiques, 1980, Maloine, Paris, 101-113.

43. Y. Bouligand, La Vie des Sciences, C. R. Acad. Sci., Paris, 1985, 2, 121-140.

44. Y. Bouligand, in Physics of Defects, 1981, Balian et al. eds., Les Houches Session 35, North Holland Publ. Co., 780-811.

45. F. Grandjean, Complete Acarological Work, 1972-1976, vol.1-7, L. van der Hammen ed., W. Junk, B.V. Publ., Antiquariaat. The Hague.

46 Y. Bouligand, in Ontogenèse et Evolution, 1986, Coll. Intern. CNRS, Dijon.

# CALCULATION OF PROTEIN STRUCTURES
# FROM NMR DATA

Werner Braun

Institut für Molekularbiologie u. Biophysik, ETH Zürich

CH-8093 Zürich,Switzerland

## Introduction

Invention of two-dimensional NMR spectroscopy ([1]-[4]), the sequential resonance assignment technique ([5]-[7]) and the development of new computer algorithms ([8]-[12]) proved to be a powerful tool to determine the spatial structure of polypeptides and small proteins in solution in atomic details ([13]-[19]). In the classic method for the determination of protein structures in single crystals by X-ray diffraction, structural calculations have been an essential step [20] in the structural interpretation of the electron density maps. In contrast calculation of polypeptide and protein structures directly from NMR data were met with high scepticism. The reason for this scepticism is the absence of a direct relation between NMR data and spatial structure as in the case of the X-ray diffraction pattern. Vicinal coupling constants [21] and nuclear Overhauser effects [22] have a direct geometric meaning for torsion angles and proton proton distances but the short range nature of these data and the fact that the observed NMR parameters are average values made it doubtful that these geometric data allow it to deduce the spatial structure of a protein directly from the experimental data.

Besides the interpretation problem of the observed NMR parameters in terms of useful and reliable geometric constraints one has to solve the difficult computational problem of determining tertiary protein structures which are compatible with the given experimental data and the stereochemical constraints. This numerical problem is related but not equivalent to the triangularization problem which consists in converting distances into 3-dimensional coordinates, because the distance information is given in the form of rather loose upper and lower bounds. In addition there are dihedral angle constraints and inherent chirality constraints for the individual amino acid residues.

The short-range character of the torsion angle and the proton proton distance information makes it also a hard problem to "guess" at the outset of the data analysis a rough global structure which could be used as a starting conformation for fitting the structure to the data. The short-range character of the data has the following meaning. In case of the vicinal coupling constants, the information on the torsion angles is of short range relative to the covalent structure, so it is straightforward to characterize a consistent local conformation in terms of torsional angles. However, the accumulation of local errors along the polypeptide chain can lead to rather large errors in the global fold. In contrast NOE data are information on short spatial distances. In proteins only proton-proton spins separated by ca. 5 Å or less give rise to a detectable NOE signal. Due to the dense packing of globular proteins short contacts between protons separated far along the polypeptide chain should abound. Quantification of this information in terms of exact distances is difficult due to internal flexibility of a protein. However, setting an upper limit to the observed short proton-proton distance is quite reliable. The calculational problem is then to convert this information from the distance space into the 3-dimensional cartesian space.

One approach is based on the metric matrix method ([8]-[11]). In the application of this method ([13],[16]) distances are converted into three-dimensional cartesian coordinates by a partial diagonalization of a certain matrix, the metric matrix. The basic equations used in this approach are directly applicable in cases when all distances between all atoms in a protein are exactly known. For data set as arising in practice the basic equations represent an approximation and several heuristic reasoning has been introduced to improve convergence and to be able to handle large distance matrices. A second method, the variable target function method [12] has been recently succesfully applied to determine the tertiary structure of several polypeptides ([14],[19]) and proteins ([17],[18]) from NMR data sets. The basic principles of both methods will be reviewed, current applications described and future developments sketched.

### 1.Formulation of the mathematical problem

Before we can proceed to formulate the mathematical problem which is to be solved in the direct method of protein structure determination from NMR data, we have to characterize the geometric constraints available from the experiments. A profound description of the type of 2D-NMR experiments needed for the collection of data in biopolymers can be found in [23].

Cross relaxation rates $\sigma_{ij}$ between two proton spins i and j can be measured by recording the intensity of NOE cross peaks as a function of the mixing time, the build up curves. The initial slope of this curve is directly related to the cross relaxation rate $\sigma_{ij}$ [24]. Recording 2D NMR NOESY spectra ([2],[3]) of small proteins nowadays can give about 500 to 1000 cross relaxation rates for individual proton proton pairs. By a multiparameter fitting method the cross relaxation rates can also be derived by a more rigorous yet more elaborate procedure from the ensemble of all build up curves ([25],[26]).

The crossrelaxation rates $\sigma_{ij}$ are given by

$$\sigma_{ij} = f(\tau_{ij}) \langle r_{ij}^{-6} \rangle \tag{1.1}$$

where $r_{ij}$ is the distance between spins i and j, and $f(\tau_{ij})$ is a function of the correlation time $\tau_{ij}$ for the reorientation of the vector connecting the two spins and the bracket $\langle\rangle$ denotes averaging over the ensemble of molecular structures interconverting in thermal equilibrium.

In a rigid protein structure the correlation time $\tau_{ij}$ between all the different pairs of protons would be identical and equal to the correlation time $\tau_R$ for the overall tumbling of the molecule. Also the thermal averaging would be trivial and eq.(1.1) could be used to calculate unknown distances $r_{ij}$ from a set of known distances $r_{kl}$ by

$$r_{ij} = r_{kl} \left[\frac{\sigma_{kl}}{\sigma_{ij}}\right]^{1/6} \tag{1.2}$$

Inherent flexibility of protein structures can be taken only crudely into account. The ratio of an effective cross-relaxation rate in a flexible protein compared to a calibration cross relaxation rate between spins with a fixed, known distance can be estimated by a function of the maximal distance $R_m$ [9]. The "maximal" distance is generally defined as the distance up to which a significant fraction,e.g 95% of the population, is occupied.

$$\frac{\sigma_{ij}}{\sigma_0} \leq Q(R_m) \tag{1.3}$$

The function $Q(R_m)$ has been determined by assuming an uniform distribution of the distances between the absolute minimal distance, the sum of the van der Waal radii of two hydrogens, and the maximal distance $R_m$. The estimate (1.3) can now be used to determine $R_m$ by measuring $\sigma_{ij}$ . This procedure has been used in [13] . Distance constraints for protons separated by at most 3 torsion angles about single bonds were treated differently from distance constraints for protons separated by more than 3 torsion angles. In the first case the rigid model was applied with 4 classes of distance limits 2.4Å, 2.7Å, 3.1Åand 4.0Å . In the second case the uniform averaging model was applied with the same levels of intensities and mixing times but loosened upper limits. In subsequent protein structure determinations, a similar scheme for the translation of NOE cross peaks into upper limit distance constraints was used ([16]-[18]).

The main conclusion is that NMR data in proteins give upper limit distance constraints or imprecise distance information with errors comparable to the size of the distances itself. On the other hand the number of distance constraints is much larger than the number of degrees of freedom.

Vicinal proton-proton coupling is another source of useful geometric information. The dependence of the vicinal coupling constant between two protons $H^1$ and $H^2$ on the dihedral angle $\theta$ is given by a Karplus type equation [21] .

$$J_{H^1 H^2}(\phi) = A + B cos\phi + C cos2\phi \tag{1.4}$$

The parameters A,B and C for the vicinal coupling constants $^3J_{\alpha NH}$ and $^3J_{\alpha\beta}$ for polypeptides have been empirically determined by a best fit procedure for the measured vicinal coupling constants for systems where also a highly refined X-ray structure was available. Numerous attempts have been done along these lines to determine the "best" set of parameters. This empirical approach assumes that X-ray and solution structures are the same and the average process is not severe. The measured values of the coupling constants are averaged over the ensemble of equilibrium conformations. This fact requires to use only the extreme values of the vicinal coupling constants for structural interpretation, because for these extreme values averaging should not have a major effect. But using only the extreme values of the Karplus curve of the vicinal coupling constant leads to a rather large inaccuracy in the dihedral angle obtained from the measured coupling constant.

All these considerations on the flexibility of the molecule lead to a similar conclusion as to which type of dihedral angle constraints can be expected in a direct method approach. As in the case of distance constraints, the experiments define an allowed interval for dihedral angles and the problem consists of finding all molecular conformers with dihedral angles in these allowed intervals.

2.Computational Methods

(a) Metric Matrix Methods

The metric matrix

$$G_{ij} = \mathbf{r}_i \cdot \mathbf{r}_j \tag{2.1}$$

where $\mathbf{r}_i$ denotes the cartesian coordinates of atom i and $\cdot$ ,the dot product, can be used to convert distances into cartesian coordinates [8]. This method gives an exact solution in cases when all distances between all pairs of atoms are known exactly. $G_{ij}$ is a $N \times N$ matrix. The matrix elements of the metric matrix determine the coordinates uniquely except for a rotation and inversion. The relation is simply given by diagonalization of the matrix.

$$G_{ij} = \sum_{\alpha}^{N} \lambda_{\alpha} E_{i,\alpha} E_{j,\alpha} \tag{2.2}$$

This relation can be seen by proving the two important properties of the metric matrix. The metric matrix is positive semidefinite and has rank three. This means that all eigenvalues of the metric matrix are greater than or equal to zero and at most three eigenvalues are different from zero. This can be derived from the quadratic form of the metric matrix.

$$\sum_{i,j}^{N} g_{ij} z_i z_j = \left( \sum_{i}^{N} z_i \mathbf{r}_i \right) \cdot \left( \sum_{i}^{N} z_i \mathbf{r}_i \right) \geq 0 \tag{2.3}$$

If the quadratic form is zero, one obtains a 3-dimensional vector equation or three linear equations in the N variables $z_i$ .

$$\sum_{i}^{N} z_i \mathbf{r}_i = 0 \tag{2.4}$$

Therefore there are at least N-3 linear independent nontrivial solutions which means that the metric matrix has at most only three eigenvalues different from zero. In the general eigenvector decomposition eq(2.2) of a metric matrix corresponding to a a set of 3-dimensional coordinates all but three terms vanish and a comparison of (2.1) and (2.2) leads to

$$r_i^{\alpha} = \sqrt{\lambda_{\alpha}} E_{i,\alpha} \tag{2.5}$$

The metric matrix can be calculated directly from the distances. This makes this quantity important for practical use.

$$G_{ii} = \frac{1}{N} \sum_{j}^{N} D_{ij}^2 - \frac{1}{2N^2} \sum_{j,k}^{N} D_{jk}^2 \tag{2.6}$$

$$G_{ij} = \frac{1}{2}(G_{ii} + G_{jj} - D_{ij}^2) \tag{2.7}$$

In the first of these equations for the diagonal term it is implicitly assumed that the structure is centered to the origin. Another choice would be setting one particular atom usually numbered 0 at the origin.

$$G_{ii} = \mathbf{r}_i \cdot \mathbf{r}_i = D_{0,i}^2 \tag{2.8}$$

As we have seen in the previous section, distance information is given in the form of an interval

$$L_{ij} \leq D_{ij} \leq U_{ij} \tag{2.9}$$

and usually only for a small subset of all possible atom pairs. It is a subset of all possible hydrogen atom pairs. The best data sets accumulated for small proteins in practice are of the order of 500 to 1000 distance constraints. This is a small number compared to all possible atom pair distances (500 000) to generate a full metric matrix.

Initial distances are chosen at random between the limits given in (2.9). This usually leads to a distance matrix not embeddable in three 3 dimensions, i.e. the metric matrix calculated from the distances is not positive semidefinite with rank three. This means there there are no coordinates in 3 dimension with the same distances as the randomly chosen distances. Approximate coordinates are then calculated by using the three greatest eigenvalues in equation (2.5).

For use in the tertiary structure determination of a polypeptide chain it is not sufficient to have an embed algorithm; one also has to combine the standard geometry of the individual amino acids (e.g. ECEPP geometric parameters [27]) in a library with the embed procedure to extract the distance constraints which define the stereo chemistry.

This was first done in the calculation of micelle bound glucagon [9] . A new FORTRAN computer program, based on the metric matrix approach [8] was written to model the geometry of the individual amino acids by distances alone. In this distance geometry approach as a model building algorithm for polypeptide chains, exprimental distance information could easily be included as additional constraint. In designing the program we had to define the standard bond lengths and bond angles by distances. This was done by interfacing the metric matrix approach with the standard amino acid library of ECEPP [27]. The only necessary input information for the chemical structure consisted then of the amino acid sequence. All relevant distance information was automatically read from the ECEPP library.

Test calculations also showed that individual chirality terms must be added to the error function in the refinement procedure to get the correct chirality of the asymmetric $C^\alpha$carbon atoms of the amino acid residues and the chirality of the $C^\beta$atom of Thr and Ile. This requirements extended the original scheme of the metric matrix approach from a pure distance geometry problem towards a refinement problem.

An extension of the approach along this lines was done in the program DISGEO [11] where several new features were included to make this approach workable also for small proteins in the pseudo atom representation. Embedding is done in two steps where first the conformation of a substructure consisting only of a subset of a third of the atoms in the complete structure is calculated and then the distances extracted from the calculated substructures are relaxed somewhat and included as additional constraints for the embedding of the complete structure.

A similar program for modeling DNA structures by distances and chirality constraints have been recently developped and successfully applied to experimental distance constraints [28].

Current research ([29]-[30]) is concerned with the correction or prediction of the undefined distances such that the complete distance matrix is embeddable. Theorems on necessary and sufficient conditions on the embeddability of the distances in the 3-dimensional space exists. The problem is to change the distances within the allowed range such that they are embeddable. No theoretical basis exists for this question. The proposed approach [30] uses Caley-Menger coordinates to fill out a complete distance matrix from a sparse incompletly defined distance matrix. This was done by a suitable simplification of the Caley-Menger determinants such that they can be calculated with minimal effort. It is not yet clear that such an approach would be applicable for a system of the size of a protein with all atoms including hydrogen atoms.

(b) Variable Target Function Method

A quite different approach, the variable target function method [12], is based on the general frame of nonlinear optimization. Torsion angles are used as independent variables to keep the number of independent variables as small as possible. The program DISMAN which implements the variable target function method generates coordinates from the dihedral angles and changes the dihedral angles in such a way that a target function becomes zero for a structure which fulfills all constraints. The target function is a measure of how good the distance constraints are fulfilled . There are many ways to construct such target functions.

A typical form of the target function is given by

$$T = \sum_{i<j} \left[ \theta(D_{ij} - U_{ij})(D_{ij} - U_{ij}) + \theta(L_{ij} - D_{ij})(L_{ij} - D_{ij}) \right] \tag{2.10}$$

The function $\theta(x)$ which is 0 for $x \leq 0$ and 1 for $x > 0$ is used to sum up all distance violations. The summation over the atompairs i and j is, of course, only over those pairs where there are constraints. The function T is 0 for a solution, is positive for all conformations not satisfying the constraints perfectly and increases as the distance constraints violations are getting worse. Usually the target functions are variations of the type (2.10) that only the square of distances is used because of efficient computation and such that they are are also continuously differentiable at the boundaries $D_{ij} = L_{ij}$ and $D_{ij} = U_{ij}$. This can be done by taking some powers of the distance violations.

Explicit restrictions on torsional angles from spin-spin coupling constants ([21],[31],[32]) can be implemented easily. This is done in DISMAN following the same philosophy used in constructing the target function from distance constraints. For each restricted torsion angle to an allowed region, i.e. to a region compatible with the NMR data, the target function is defined as zero within the allowed region, has continuous first derivatives at both region boundaries and increases smoothly with the amount of deviation from the allowed region. Some care has to be taken, because the 3 dimensional coordinates are $2\pi$ periodic functions of the torsion angles. Also the evaluation of the function should be fast. In DISMAN a fourth order polynomial with continous first derivatives and $2\pi$ periodicity is constructed.

The method of variable target functions means that one does not try to minimize T at once but rather to minimize gradually a series of functions which approximate T. More specifically, for a polypeptide chain of n residues the target functions $T_{k,l}$   $k = 1, 2, \ldots, n$ and $l = 1, 2, \ldots, n$ only include those terms of the form as in (2.10) for atom pairs belonging to residues with difference of their sequence numbers less than k if the upper or lower limits are from NMR data or less than l if the lower limits are the sum of repulsive core radii. The strategy consists in first minimizing $T_{k,l}$ with small values of k and l and then gradually increasing k and l up to n. The final solution of the problem consists of one or several conformations having zero values for $T_{n,n} = T$ . The exact definition of the terms used in DISMAN can be found in [12]. In case of an overdetermined problem the best conformation consistent with the input distance information and stereo-chemical criteria is the one which gives the global minimum of the target function.

This strategy was shown to be effective if good distance information of a short range nature is available. Exact characterization of good distance information which can guide the conformation from correct short range to medium or long range conformations is missing. More extensive numerical experience is certainly needed. Some heuristic ideas of describing the success of the method are as follows. Short and long range distance constraints impose different type of restrictions on the polypeptide conformation . Once short range distance constraints are fulfilled, the polypeptide chain keeps a large amount of "flexibility" for those conformational changes maintaining the short range distance information. Small local changes can give rise to drastic global changes. So these small changes can be used to satisfy the long range distance constraints.

The variable target function method certainly has some resemblance to restrained energy minimization in torsion angle space ([33],[34]). Two new features are an efficient way to calculate gradient information of the target function ([35],[36]) and the way the target function in (2.10) is minimized. In the variable target function method a series of target functions $T_{k,l}$ is minimized rather than a certain pseudo energy function. This approaches the local minima problem in a quite different way than modification of the infinite repulsion energy terms by finite models for overlap atoms (' soft atoms'). This means that in the stage of the minimization of $T_{k,l}$ all atom pairs whose residue numbers differ by more than l can freely penetrate each other, whereas there is still some barrier in the soft atom model. Also the restraints or constraints are brought differently into play by the two methods.

### 3. Applications

(a) X-ray structure of BPTI as a test protein

Test calculations usually are done by extracting distances from a given X-ray structure and setting the upper or lower limits to some range near the extracted distances. In the calculations with these simulated distance constraints sets ([12],[37]) one first wants to demonstrate that for a sufficient complete distance constraints data set, the calculated structures converge to the structure from which the distances were extracted. Also one wants to explore the theoretical structural consequences of distance data sets to set guidelines for the experimental work. What structural features can be expected in a typical experimental data set? Which additional data can significantly improve the structures ? Present calculations already indicate where improvements of the structures can be expected.

As an illustrative example of the type of calculations, the results of the DISMAN calculations with several distance constraints data sets of BPTI are presented [12]. In test calculations of this type one has to separate the influence of the starting conformations from the influence of the data set. The approach taken in this study was therefore first to generate 10 structures by choosing the variable dihedral angles randomly.

A standard measure to quantify the differences of two molecular conformations is the root mean square distance (r.m.s.d) [39] . It gives roughly the average distance between all equivalent pairs of atoms after the two structures have been optimally superposed. Values of 10 Å to 20 Å for small proteins mean that the two structures are clearly globally different. Values around 2 Å to 3 Å are typical for structures with the same global fold but several local deformations. Molecular mechanics calculations of several small proteins show that the thermal fluctuations are in the order of 0.5 Å to 1 Å [40].

The r.m.s.d values comparing all pairs of initial structures ranged from around 8 Å to 22 Å . This set of initial structures were then used as starting conformation with several data sets. By using the most stringent data set (EX5) where all exact short proton-proton distances less than 5 Å have been used as constraints, the polypeptide backbone of BPTI were nearly exactly regenerated with r.m.s.d values ranging from 0.01 Å to 0.15 Å by starting from the 10 randomly chosen initial structures. This again shows that short proton-proton distances are potentially a quite powerful source of information for restricting the global polypeptide fold.

Data sets of the type experimentally available by the present NMR techniques are the data sets AL5 and AU5 defined below. Both data sets consist of short and long range constraints. Short range distance constraints were defined as constraints between the protons NH, $H^{\alpha}$ and $H^{\beta}$ which belong to residues separated sequentially by 2 or less intervening residues. Intraresidue constraints were deliberately excluded, because the stereospecific assignment of the methylene protons, a tedious and difficult procedure, for all residues is required. These distance types were put into 6 classes from 2 Å to 5 Å with an interval of 0.5 Å . The upper and lower bound distance constraints were defined as the upper and lower bounds of the class to which it belongs. In AU5 only the upper bounds for the short range constraints were used, in AL5 upper and lower bounds were included. For the long range constraints upper limit distance constraints were set in both data sets to 5 Å if the corresponding proton-proton distance were less than 5 Å . So the two data sets differ in the short range data sets where in AL5 also lower limits have been included. Since quantification of lower limits from the NOE data is difficult because of the inherent flexibility of the protein, we want to study the theoretical implications for the structure by this data set.

Both data sets were used in calculations with the same initial random structures. Because of restricted computer time we had to choose 3 among the 10 previously generated initial conformations. The average r.m.s.d values comparing the calculated structures with the BPTI X-ray structure for the data set AL5 were 1.4 Å and 2.3 Å for backbone and all atoms, respectively, and 1.5 Å and 2.5 Å for AU5. The result indicates that the differences in the restrictions of structures between the data sets AL5 and AU5 is not significant. Comparing the calculated structures to the X-ray structure, one recognizes well defined side chain conformations in the interior of the protein, especially the orientation of the aromatic ring planes. This is not a trivial consequence of the data sets, because the long range distance constraints were chosen in both data sets with a rather loose upper bound of 5 Å . This indicates that part of this restriction is certainly due to the packing constraints in the interior of the globular protein.

(b) Experimental data of BPTI

Over several years an extensive set of distance and torsion angle constraints have been compiled for basic pancreatic trypsin inhibitor(BPTI) , the favorite test protein for several experimental and theoretical studies which include the distance geometry calculations with simulated NMR data sets. BPTI also played a pilot role in the development of several of the methods described with the structure determinations of other proteins.

In the calculation with the experimental BPTI data, care was taken that the two programs DISGEO and DISMAN used exactly the same data sets. So the results obtained from that study give indications on the sampling property of the two programs. The NMR structures obtained by two different programs should also be a reliable basis for comparing the highly refined X-ray structure [38] in single crystals to the solution structure.

Comparison of the final distance and torsion angles violations of the structures calculated by both programs shows that the quality of convergence obtained with the two programs is nearly the same. No structures were found where all input constraints were exactly satisfied but the final distance and torsion angle violations are tolerable if we consider the accuracy of the estimations of these data. This also showed that there are no severe inconsistencies in the NMR data set and the problem of averaging over two very different states is certainly not a major one in the direct determination of the tertiary protein structures.

Root mean square distance values (r.m.s.d) [39] measure the deviations of the backbone and side chain conformations of the NMR structures to the X-ray structure. Both programs give quite similar values if one compares the NMR to the X-ray structure, e.g. 2.4 Å and 2.3 Å for backbone atoms for DISMAN and DISGEO, respectively, and 2.7 Å for both programs if in addition the constrained side chain atoms were included. Side chains were considered as constrained when there was at least one distance constraint on a side chain atom. In the case of the long residues Lys and Arg the constraint must be beyond the $\beta$ proton. Differences of the behaviour of the two programs can be seen by calculating the r.m.s.d values within the set of structures calculated by each program. The variability among all DISGEO structures is significantly smaller than among the DISMAN structures. This is particular pronounced in the average r.m.s.d values of the non-constrained side chains( 4.2 Å versus 2.4 Å for DISMAN and DISGEO). This indicates that the DISGEO program has a tendency to underestimate the true variations of the structures compatible with the data.

The study also again showed that the global polypeptide fold could be reliably determined by the NMR data whereas the local structures can still be improved significantly. The average standard deviation for backbone dihedral angles averaged over all residues is still quite high in both programs (around 60°). On well defined segments the individual deviations are, however, in the range of 20° to 30°. This result is not a problem of convergence but a property of the NMR data sets.

(c) Metallothionein-2

Metallothioneins are small, metal- and cystein rich proteins. Metal storage or heavy metal detoxification were proposed as physiological functions for these proteins [41].

Preliminary calculations with the DISMAN program to determine the global polypeptide fold of this protein with the NMR data described below were promising [17]. This study also showed that the existence of some regular secondary structural elements which could be identified by typical NOE patterns [42]. is not necessary for protein structure determination by NMR data. At the beginning of this study no X-ray structure of any metallothionein was available. At present a 2.3 Å resolution X-ray structure of rat liver Cd,Zn metallothionein [43] can be used to study similarities and differences between single crystal and solution structures.

The NMR information was translated into distance constraints as input for the DISMAN program in the following way. The metal cysteine connectivities, combined with the amino acid position of the sequence specific assigned cysteines provide a dense network of distance constraints for the two clusters containing 3 and 4 cadmium metal ions. A tetrahedral arrangement of four sulfur atoms around each metal ion, with a cadmium-sulfur distance of 2.6 Å , was assumed. This assumption is based on several spectroscopic experiments. The seven Cd ions were thereby represented by 20 pseudo atoms covalently linked to the sulfur atom of each of 20 metal bound cysteines. Equivalent Cd reference points were then forced to coincide by distance constraints and in the final structures the real Cd ion positions were calculated as the average positions of the equivalent Cd reference points.

Large scale calculations are currently underway with a more comprehensive set of $^1H - ^1H$ distance constraints from NOESY spectra recorded with different mixing times and numerous torsion angle constraints for $\phi$ and $\chi^1$ by combined evaluation of intraresidue, sequential NOE's and the vicinal coupling constants $^3J_{\alpha NH}$ and $^3J_{\alpha\beta}$ for rabbit and rat liver $Cd_7$ MT2. The extensive calculations also should give a larger ensemble of final structures to improve the statistical significance of the r.m.s.d values. The NMR structures obtained should give a firm basis for comparing them to the X-ray structure. It is known there are a quite large number of different metal-cysteine coordinations found by the two techniques.

(d) $\alpha$-Amylase inhibitor

$\alpha$-Amylase inhibitor from Streptomyces tendae, tendamistat binds tightly to and inhibits specifically mammalian $\alpha$-amylases [44]. The polypeptide chain consists of 74 amino acid residues, among them 4 Cys residues forming two disulfide bridges(Cys11-Cys27,Cys45-Cys73). An extensive set of constraints from NMR experiments could be assembled and used in the structure determination [18]. These constraints include 401 distance constraints from nuclear Overhauser effects, 168 distance constraints from hydrogen bonds and disulfide bridges, and 50 torsion angle constraints from measurements of spin-spin coupling constants. Calculations with the DISMAN program provided four structures compatible with the NMR constraints. The initial structures were randomly chosen with no assumption on the existence of regular secondary structures.

Parameters to judge the quality of the obtained structures are the residual violations of the NMR constraints. The average violation of the distance constraints per constraint dropped from the range 10 Å to 15 Å for the initial structures to about 0.025 Å for the final structures, indicating that all the final structures were in agreement with the NMR constraints and were entirely determined by the NMR data and not by the initial conformations.

A second numeric quantity is the r.m.s.d value for the backbone atoms; averaged over all pairwise comparisons between the final structures, this was 1.6 Å for residues 6 to 73. This compares with 15.2 Å for the average value of the backbone atoms in the same residue region of the initial conformation.

Other more qualitative criteria are the dense packing in the interior of globular proteins and handedness properties as found to be typical for globular proteins . One is the right-handed twist of the $\beta$-sheet structure which could be checked in the calculated structures.

On the proposal of R. Huber (MPI Munich), the structural analysis of $\alpha$-amylase inhibitor was done independently and in parallel by X-ray diffraction methods and NMR techniques. This should provide an objective test for the reliability of the NMR structure determination and to make sure that differences and similarities between NMR and X-ray structures are not biased by the exchange of structural information between the two groups using the two different methods. In the future a combination of both techniques and the optimal use of the structural results of either method might be used to speed up tertiary structure determination of proteins.

The global polypeptide fold of the NMR structures [18] and of the X-ray structure [45] of $\alpha$ amylase inhibitor closely coincide [46]. The structures form a Greek-key $\beta$-barrel with the topology +1,+3,-1,-1,+3 . The coincidence of the fold of the polypeptide chain as found by the two methods is especially close in those segments which are narrowly confined by NMR data. These segments include the 6 strands of the two $\beta$-sheets. Deviations are pronounced at less well defined regions such as the segment from residues 62 to 66. The side chain conformations of the segment Trp18-Arg19-Tyr20 which are presumably involved in binding to the amylase, can be described as a sandwich structure found by both methods. This is a remarkable result because the residues of this segment form part of the surface.

4.Conclusion

Use of 2D NMR techniques and distance geometry calculations has been proven to be a reliable and practical tool to determine the tertiary structure of small proteins in solution. Differences between X-ray structure in single crystals and NMR structure in solution can now be studied in atomic details. Combined use of the X-ray and NMR technique might lead to an accelerated determination of biological important proteins.

REFERENCES

[1] J. Jeener, B.H. Meier,P. Bachmann & R.R. Ernst, J. Chem. Phys. 71,4546 (1979)

[2] S.Macura, & R.R. Ernst, Mol. Phys. 41,95 (1980)

[3] Anil Kumar, G.Wagner, R.R Ernst & K. Wüthrich, J. Amer. Chem. Soc. 103,3654 (1981)

[4] K. Nagayama & K Wüthrich, Eur. J. Biochem. 115,653 (1981)

[5] K. Wüthrich, G. Wider, G. Wagner, W. Braun, J. Mol. Biol. 155,311 (1982)

[6] M. Billeter, W.Braun & K. Wüthrich, J. Mol. Biol. 155,321 (1982)

[7] G. Wagner & K. Wüthrich, J. Mol. Biol. 155,347 (1982)

[8] G.M. Crippen & T.F. Havel, Acta Crystallogr. A 34,282 (1978)

[9] W. Braun, C. Bösch, L.R. Brown, N. Gō & K. Wüthrich, Biochim. Biophys. Acta 667,377 (1981)

[10] T.F. Havel, I.W. Kuntz, & G.M. Crippen, Bull. Math. Biol. 45,665 (1983)

[11] T.F. Havel, & K. Wüthrich, Bull. Math. Biol 46,673 (1984)

[12] W. Braun & N. Gō, J. Mol. Biol. 186,611 (1985)

[13] W. Braun, G. Wider, K.H. Lee & K. Wüthrich, J. Mol. Biol. 169,921 (1983)

[14] Y. Kobayashi, T. Ohkubo, Y. Kyogoku, Y. Nishiuchi, S. Sakakibara, W. Braun & N. Gō, Proc. 9th Am. Peptide Symp.(K.D.Kopple & C.M. Deber, ed.) Pierce Chem. Comp.

Rockford (1985)

[15] R. Kaptein, E.R.P. Zuiderweg, R.M. Scheek, R. Boelens & W.F. van Gunsteren, J. Mol. Biol. 182,179 (1985)

[16] M.P. Williamson, T.F. Havel & K. Wüthrich, J. Mol. Biol. 182,295 (1985)

[17] W. Braun, G. Wagner, E. Wörgötter, M. Vasak, J.H.R. Kägi, & K. Wüthrich, J. Mol. Biol. 187,125 (1986)

[18] A.D. Kline, W. Braun, & K. Wüthrich, J. Mol. Biol. 189,377 (1986)

[19] T. Ohkubo , Y. Kobayashi, Y. Shimonishi, Y. Kyogoku, W. Braun, & N. Gō, Biopolymers 25,123 (1986)

[20] T.L. Blundell & L.N. Johnson, Protein Crystallography, Academic Press,New York (1976)

[21] M. Karplus, J. Chem. Phys. 30, 11 (1959)

[22] J.H. Noggle & R.E. Schirmer, The Nuclear Overhauser Effect, Academic Press, New York (1971)

[23] K. Wüthrich, NMR of proteins and nucleid acids, Wiley , New York (1986)

[24] G. Wagner & K. Wüthrich, J. magn. Reson. 33,675 (1979)

[25] J.W. Keepers & T.L. James, J. magn. Reson. 57,404 (1984)

[26] E.T. Olejniczak, R. T. Gampe & S.W. Fesik, J. magn. Reson. 67,28 (1986)

[27] F.A. Momany, R.F. Mc GUIRE, A.W. Burgess & H.A. Scheraga, J. Phys. Chem. 79,2361 (1975)

[28] D. Hare, L. Shapiro & D.J. Patel, Biochemistry, 25,7445 (1986)

[29] M.J. Sippl & H.A. Scheraga, Proc. natn. Acad. Sci. U.S.A. 82,2197 (1985)

[30] M.J. Sippl & H.A. Scheraga, Proc. natn. Acad. Sci. U.S.A. 83,2283 (1986)

[31] A. DeMarco, M. Llinas & K. Wüthrich, Biopolymers 17,617 (1978)

[32] A. DeMarco, M. Llinas & K. W"uthrich, Biopolymers 17,637 (1978)

[33] G. Nemethy & H.A. Scheraga, Quart. Rev. Biophys. 10,239 (1977)

[34] M. Levitt, J. Mol. Biol. 170,723 (1983)

[35] T. Noguti & N. Gō, J. Phys. Soc. (Japan) 52,3685 (1983)

[36] H. Abe, W. Braun, T. Noguti & N. Gō, Computers & Chemistry 8,239 (1984)

[37] T.F. Havel & K. Wüthrich, J. Mol. Biol. 182,281 (1985)

[38] J. Walter & R. Huber, J. Mol. Biol. 167,911 (1983)

[39] A.D. McLachlan, J. Mol. Biol. 128,49 (1979)

[40] M. Karplus & J.A. McCammon, C.R.C. Crit. Rev. Biochem. 9,293 (1981)

[41] J.H.R. Kägi & M. Nordberg, Metallothionein, Birkhäuser-Verlag, Basel (1979)

[42] K. Wüthrich, M. Billeter & W. Braun, J. Mol. Biol. 180,715 (1984)

[43] W.F. Furey, A.H. Robbins, L.L. Clancy, D.R. Winge, B.C. Wang & C.D. Stout, Science 231,704 (1986)

[44] L. Vertesy, V. Oeding, R. Bender, K. Zepf & G. Nesemann, Eur. J. Biochem. 141,505 (1984)

[45] J.W. Pflugrath, G. Wiegand, R. Huber & L. Vertesy, J. Mol. Biol. 189,383 (1986)

[46] W. Kabsch & P. Rösch, Nature 321,No 6069,469 (1986)

# STRUCTURAL ANALYSIS AT MOLECULAR DIMENSIONS OF PROTEINS AND PROTEIN ASSEMBLIES USING ELECTRON MICROSCOPY (EM) AND IMAGE PROCESSING

**Ueli Aebi**

M.E. Müller-Institute for High Resolution Electron Microscopy at the Biozentrum,
University of Basel, CH-4056 Basel, Switzerland
and
Department of Cell Biology & Anatomy, The Johns Hopkins University
School of Medicine, Baltimore, MD. 21205, USA

## Introduction

The major advantage of EM over X-ray diffraction is its ability to directly record an **image** (i.e. amplitudes and phases) and not just a **diffraction pattern** (i.e. amplitudes only) of the specimen under investigation. The disadvantage of EM, however, is the fact that, despite the **near-atomic** (i.e. 2-3 Å) resolution performance of a state-of-the-art instrument, the **practical** resolution with most biological specimens is typically one order of magnitude worse (i.e. 20-30 Å). This is primarily due to the following specimen-dependent limitations: (1) biological (i.e. carbonaceous) material has relatively **low inherent contrast** in an (i.e. 100 kV) electron beam; (2) the alterations (e.g. denaturation and collapse) accompanying the preparation (e.g. dehydration) of biological material for inspection in the high vacuum of an EM cause serious **specimen preparation artifacts**; (3) biological material is extremely **radiation sensitive** when bombarded with electrons [c.f. 1-3].

One or several of the above mentioned limitations may partially be overcome under favourable conditions and/or with a "cooperative" specimen such that in a few cases significant structural detail on the 5-10 Å resolution level has been obtained [c.f. 4-7]. Hence, in most cases, the EM has not provided us with **absolute** (i.e. near-atomic) structure - which is typical with X-ray diffraction analysis - but rather with a **relative** (i.e. the overall size and shape and, possibly, a "surface envelope") representation of the protein molecule under investigation. Furthermore, it should be emphasized that, despite the fact that the EM is capable to directly provide us with images of protein molecules, to reveal statistically significant high resolution (i.e. <20 Å) information from them, we have to **average** information from many molecular images after we have aligned them relative to each other to improve the **signal-to-noise (S/N)** ratio of the generally low-contrast/high-noise molecular images. Therefore, it is most convenient to have available ordered arrays (e.g. two-dimensional (2-D) or helical arrays) of the protein molecule to be investigated  since those provide us with a **rule** as to how to translate and/or rotate individual molecules (i.e. "unit cells") within the array to bring them into register for averaging. Sometimes such ordered protein arrays or **"bio-crystals"** occur **in vivo**  (e.g. virus capsids, filaments, cell surface protein layers, membrane protein patches, etc; for examples see [8]), but more often they have to be induced **in vitro** from purified protein by establishing appropriate polymerization or crystallization conditions [c.f. 8].

### Contrast Enhancement

There are basically two avenues which may be pursued when attempting to improve the  low inherent contrast of biological material: (1) **instrumental** contrast enhancement; (2) **specimen** contrast enhancement (see section on "Specimen Preparation").

In the case of instrumental contrast enhancement we have the following possibilities when forming an image with the electrons (e$^-$) which have traversed the specimen and therefore either being **unscattered, elastically scattered**, or **inelastically scattered**:

(1) **Aperture contrast**: In the case of a thin (i.e. $\leq$500 Å) biological specimen, about 25-50% of the e$^-$ which have interacted with the specimen are elastically scattered and therefore are deflected by relatively large angles (i.e. 10-150 mrad) from their incident direction. By placing a circular aperture in the back-focal plane of the imaging lens, a significant  fraction of these elastically scattered e$^-$ can be stopped from reaching the image plane, resulting in an image which is primarily formed by the unscattered and the inelastically scattered (i.e. 0-15 mrad) e$^-$. As a consequence, **thick** areas (i.e. which scatter a lot of e$^-$) of the specimen appear **dark**, while **thin** areas (i.e. which scatter few e$^-$) of the specimen appear **bright** in the image plane.

(2) **Dark field**: Instead of preventing most of the elastically scattered e$^-$ from reaching the image plane as this is done in the case of aperture contrast (see (1) above), we may exclude the inelastically scattered e$^-$ from the image by putting a "beam stop" into their path  in the diffraction plane of the imaging lens. In this case, thick specimen areas appear bright, whereas thin specimen areas appear dark in the image plane (i.e. opposite to what is observed with aperture contrast).  The most effective way to just collect the elastically scattered e$^-$ is with a **scanning transmission EM (STEM)** equipped with an **annular** - or dark field - detector, which is a ring detector with a hole in the center just big enough that the unscattered and most of the inelastically scattered e$^-$ will pass

through [c.f. 9]. Since with the STEM a finely focused (i.e. ≤10 Å diameter) e⁻ beam is stepped over the sample pixel by pixel, a **scattering experiment** is performed in each sample point of the specimen, allowing the elastically, inelastically and unscattered e⁻ to be counted separately. While the elastically scattered e⁻ are counted with the dark field detector, the unscattered and inelastically scattered e⁻ - which pass through the central hole in the dark field detector - can be run through a **spectrometer** where they are sorted according to their element specific e⁻ energy losses, thereby allowing **e⁻ energy loss spectroscopy** (**EELS**) to be performed. Hence the STEM is an **analytical** tool which enables us to measure the mass contained in a given sample volume (i.e. by counting the number of elastically scattered e⁻ coming from this specimen area ---> **mass mapping** [c.f. 10]), or which can determine the elemental composition of a given specimen area (i.e. from the characteristic e⁻ energy losses occuring in this area - which are element specific ---> **elemental mapping** [c.f. 11]). Finally, we may also form an image with just the elastically scattered e⁻ (i.e. a **dark field image**), or with just the e⁻ which have lost a characteristic amount of energy (i.e. an **electron spectroscopic image** or **ESI** [c.f. 12]).

(3) **Phase contrast**: Very much like in a light microscope, the contrast of the image in an EM may be manipulated by altering the **phase shift** between the scattered and the unscattered e⁻ waves. As shown by **Zernike**, optimal contrast is achieved when the phase shift becomes $\pi/2$ or **1/4 of a wavelength** (i.e. $\lambda/4$). While in the light microscope this phase shift is generated by a $\lambda/4$ **phase plate**, in the EM, this is achieved by a combination of the **spherical aberration** and the proper amount of **defocus** of the objective lens, parameters which, in turn, define the **contrast transfer function (CTF)** - or a **virtual** phase plate - of the objective lens of the EM [c.f. 13]. In this case, the optimal phase shift is approximately reached when the objective lens is slightly **underfocused** (i.e. the lens current is underexcited relative to the in-focus current setting). In practice, however, the exact amount of underfocusing will depend on the spatial frequencies one wants to optimally enhance. For a given underfocus setting certain spatial frequencies will be amplified while others will be attenuated or transferred with opposite sign (i.e. **contrast reversal**) to the image plane. This is because the CTF starts oscillating such that with increasing amount of underfocusing oscillation starts for increasingly smaller spatial frequencies. Therefore, care and judgement are needed to avoid loss of resolution and/or introduction of spurious structure through contrast reversal when trying to increase the overall contrast in the image by defocusing the objective lens. Nevertheless, phase or defocusing contrast is probably the most effective instrumental means to increase the image contrast of biological matter in the transmission EM.

### Specimen Preparation

To introduce biological matter - which is **wet** - into the high vacuum of an EM, it has to be **dehydrated**. This step, however, which is often achieved by simple **air-drying**, **denatures** proteins, and the **surface tension** which occurs at the air-liquid interface, causes protein molecules or supramolecular protein assemblies to **collapse** or to **spread-flatten** on the specimen support [c.f. 2,14]. Surface tension may be reduced if dehydration is performed via **freeze-drying** or **critical-point drying** of the specimen. To minimize both protein denaturation, as well as surface tension, workers in the field have developed liquid nitrogen - or even liquid helium - cooled **cold stages**, enabling them to circumvent dehydration by inspecting the specimen in the EM when embedded in a thin film of - preferentially vitreous - **ice** [c.f. 15,16]. Electron diffraction patterns recorded from **frozen-hydrated** 2-D crystalline specimens have demonstrated that with this method structural detail may be preserved to at least 3.5 Å [c.f. 17]. More recently, it has been demonstrated that with the availability of sufficiently stable cold stages, even **images** of frozen hydrated material can be recorded preserving reproducible structural detail to better than 10 Å resolution [c.f. 7].

Whether dehydrated or embedded in a thin film of ice, biological (i.e. carbonaceous) matter exhibits relatively low inherent contrast when imaged by e⁻. As a consequence, several methods have been developed to increase the inherent contrast of biological material - the two most commonly used being the following:

(1) **Negative staining**: With this method, the biological matter is dehydrated - usually by air-drying - in the presence of a **heavy metal salt** (e.g. 1% uranyl acetate or formate, 2% Na-phosphotungstate, etc). In this case, the predominant contrast is coming from the heavy metal salt **"replica"** surrounding the sample. This, however, means that we are not primarily looking at the biological material itself but rather at a **"negative"** **replica** of it. While negative staining is a quick and effective method to prepare biological material out of suspension (e.g. isolated protein molecules or supramolecular protein assemblies such as virus, filaments, 2-D or helical arrays, etc), in the best case, it enables us to map out the **overall size and shape** - i.e. in the form of a **"surface envelope"** - of a protein molecule, however, it does not allow us to look **"inside"** a protein molecule. Typically, the amount of structural detail which may **reproducibly** be resolved with negative stain lies at the 15-30 Å level. As a consequence, negatively stained biological matter will only yield a **relative** representation of protein molecules rather than their near-atomic or **absolute** structure as this is often achieved by X-ray diffraction if suitable 3-D protein crystals of the protein under investigation are available. Since, usually, the sample is more or less completely "submerged" in a "sea" of heavy metal salt, the thus obtained electron micrographs represent **through-projection** (i.e. a 2-D projection) images of the inherently 3-D specimens.

(2) **Heavy metal replication**: If we want to reveal the **surface** of our specimen, we may replicate it with a metal coat by evaporating metal (e.g. platinum or tungsten) from a certain **elevation angle** (typically 15-30⁰ relative to the specimen plane) onto the

sample surface, after the material has been dehydrated by e.g. freeze-drying. If the direction of shadowing is kept constant relative to the specimen, it is called **unidirectional shadowing**. In this case, depending on the elevation angle chosen, "ridges" will be more or less covered with metal grains, whereas "valleys" will lie in the "shadows" casted by the ridges staying in front of them relative to the metal evaporation source. While such unidirectionally shadowed metal surface replicas may sometimes puzzle the untrained eye due to the "shadow casting" effect they exhibit, they, in fact, contain useful information about the **surface topography** of the specimen which may be represented in the form of a **surface relief reconstruction** computed from the shadowgraph(s) [c.f. 18,19]. **Rotary shadowing** may be employed to generate a surface replica where the metal evaporation source is kept at a constant elevation angle while the sample is rotated relative to it. The **"rotary shadowing"** technique is often very effective when trying to image single protein molecules, particularly when combined with "glycerol spraying" the sample onto a freshly cleaved mica surface prior to air-drying/metal-shadowing the specimen at a low elevation angle (e.g. 3-10$^0$) [c.f. 20]. Sometimes, instead of revealing an evenly distributed metal coat, one may observe **preferential accumulation** of metal grains at particular locations on the specimen surface. This so-called **"decoration"** effect is due to distinct surface properties (e.g. charge, hydrophobicity, etc) of the sample which, in turn, may yield "nuclei" for accumulating metal grains.

In an attempt to preserve biological matter in a more native state and environment for EM, it has been embedded in a non-volatile **glucose syrup** [c.f. 21]. Since the density of glucose is very similar to that of protein, thus prepared specimens reveal extremely low contrast. Furthermore, glucose is very radiation sensitive itself, therefore providing no radiation protection to the highly radiation sensitive biological material. As a consequence, low (i.e. $\leq 1$ e$^-$/Å$^2$) e$^-$ doses have to be employed to preserve the high resolution structural detail in the specimen. Such low e$^-$ doses, in turn, give raise to **statistically noisy** images (i.e. with a low S/N ratio - $\leq 0.1$), thus making detection of the low-contrast structural features even more difficult. Nevertheless, when applied to highly ordered 2-D crystalline protein arrays allowing averaging over several thousand unit cells, the **averaged unit cells** of thus prepared samples may ultimately reveal reproducible structural detail in 3-D to better than 10 Å resolution, a level sufficient to visualize secondary structural features of protein molecules [c.f. 4,21].

### Radiation Damage

In addition to specimen preparation (see above), radiation damage represents the other major limiting factor for obtaining high resolution structural detail of biological matter in the EM. Typically, unprotected biological material will suffer significant damage after exposure to an accumulated dose of $\geq 1$ e$^-$/Å$^2$ [c.f. 21]. The damage is caused by the inelastically scattered e⁻ which transfer sufficient energy to the sample to **ionize atoms** and/or to **break covalent bonds**. There are basically two ways to avoid - or at least minimize - radiation damage: (1) **low dose EM**; (2) **radiation protection** of the specimen.

Apart from the technical difficulties associated with low dose EM, low-dose/high-resolution electron micrographs of unstained biological matter yield a statistically highly insignificant S/N ratio (i.e. S/N $\leq$ 0.1) for structural detail below about 10 Å resolution. This means that averaging over many (i.e. $\geq$1,000) noisy molecular images - which can translationally and/or rotationally be aligned to high accuracy - becomes a prerequisite. This, in turn, is most easily being achieved if highly ordered crystalline arrays of the protein molecule to be analyzed can be obtained. In fact, some interesting proteins already exist as **ordered supramolecular structures** (i.e. "bio-crystals") **in situ** (e.g. the "purple" membrane of halobacterium Halobium [c.f. 21], virus capsids, filamentous structures, cell surface protein layers, etc), others, however, may be induced to form such ordered arrays **in vitro** [c.f. 8].

While looking at unstained/unreplicated biological material is obviously the way to go if one wants to yield reproducible high resolution (i.e. <10 Å) structural detail about protein molecules in their **native conformation,** as yet, the most effective way to **partially protect** biological matter from radiation damage and, at the same time, **preserve a fair amount** of its structural detail (i.e. at the 15-30 Å resolution level) has, no doubt, been negative staining (see above). It should be noted, however, that even in the presence of negative stain care has to be taken with the total e⁻ dose given to the specimen since the heavy metal salts commonly used for negative staining (see above) have a tendency to crystallize and/or to migrate or redistribute over the biological material under the e⁻ beam. Nevertheless, in the presence of negative stain the specimen will resist **10-100x more e⁻** before siginificant radiation damage will become apparent.

Another promising avenue to combat radiation damage of native (i.e. unstained) biological matter has been **low temperature** EM: typically, when kept at liquid nitrogen temperature, biological material becomes **3-10x more radiation resistant** than when kept at room temperature. Even larger protection factors have been measured upon cooling the specimen to liquid helium temperature during observation in the EM.

### Image Processing

There are several reasons why image processing steps have to be applied to electron micrographs to **optimally** and **reproducibly** extract their **high resolution 3-D structural features**:

(1) **Image restoration**: The behaviour of the **contrast transfer function (CTF)** of the objective lens in the EM (see above) is such that certain spatial frequencies are amplified while others are attenuated as a function of the amount of defocus used to record the image. To compensate for this effect, as well as for residual astigmatism and other lens aberrations, image restoration steps usually have to be employed [c.f. 13,22] to faithfully represent the high resolution (i.e. <10 Å) structural features contained in such electron micrographs.

(2) **Image enhancement**: As discussed (see above), the e⁻ dose should be kept low to minimize radiation damage of the specimen, a situation giving raise to relatively high **quantum noise** in the electron micrographs. **Additional sources of noise** in the electron micrographs stem from (a) the "granularity" (i.e. the phase contrast structure) of the carbon support film, (b) granularity and irregularities of the heavy metal salt replica surrounding the biological material in the case of negatively stained samples, and (c) irregularities and/or irreproducible structural features among **supposedly identical** molecular images caused by specimen preparation artifacts and/or radiation damage. Taken together with the relatively low inherent contrast of biological matter, these various sources of noise generally result in a **statistically insignificant S/N ratio** (e.g. ≤1) of the structural detail of interest in electron micrographs recorded from biological specimens. Therefore, to arrive at a **statistically significant S/N ratio** (i.e. ≥3), image enhancement steps have to be employed which, in turn, require **averaging over a number of noisy but supposedly identical molecular images**, hence we need **structural redundancy**. As might be expected, the **alignment/orientation** steps preceding averaging become relatively simple and straightforward when having at our disposal highly ordered (e.g. 2-D or helical) arrays of the protein molecules under investigation (see above and c.f. [8]). Sometimes, however, only poorly ordered arrays - or none at all - may be obtained from the protein molecule to be studied. In this situation, **correlation** methods are now commonly used to translationally and rotationally align a set of noisy molecular images relative to a **reference image** chosen among them, followed by **single particle averaging** [c.f. 23,24].

While at first glance a set of molecular images may look identical except for the noise contaminating each image, closer inspection may yield that there exist subtle structural differences among them. By applying simple averaging techniques to them, as outlined above, inherent differences among molecules will be treated as "noise" and therefore be averaged out. During attempts to overcome these limitations, **multivariate statistical data analysis** has been employed to sort a **heterogeneous population** of particle images into **subpopulations** (or subsets) of images, such that images within a particular subpopulation are more similar to each other than to members belonging to any other subpopulation [c.f. 25,26]. As might be expected [c.f. 27], multivariate statistical data analysis applied to aligned particle images may become a powerful tool to localize specifically bound labels (e.g. antibody fragments), to sort out different projections of the same molecule, to determine magnification scaling, or to trace the fate of radiation-induced structural changes.

(3) **3-D Structure Reconstruction**: Since both stained and unstained biological matter is essentially translucent to e⁻, and since the depth of focus of e⁻ lenses is much greater than the thickness of the specimen, structural features along lines parallel to the e⁻ beam are projected on top of each other in the image. As a consequence, transmission electron micrographs represent **2-D projection images** of the inherently **3-D specimen**. Therefore, to recover the 3rd dimension, a set of **independent** 2-D projection images - i.e. viewing the 3-D specimen from different angles - have to be recorded (e.g. by **tilting** the specimen relative to the e⁻ beam), and these have to be computationally recombined by **3-D reconstruction** steps [c.f. 28,29], very much like this is performed in **computerized axial tomography** (**CAT** scanning: X-ray, PET, or NMR), a discipline which has become very powerful in modern **medical radiology and imaging**.

Although a number of methods have been proposed for 3-D structure reconstruction from a set of 2-D projections [c.f. 30], the **Fourier** method [c.f. 28] has remained the most widely used in combination with EM. It is based on the **"projection"** theorem, which states that the 2-D Fourier transform of a projection of a 3-D object is identical to a **central section** through the 3-D Fourier transform of that object. Once the 3-D Fourier transform has been **sampled** sufficiently finely by the central sections corresponding to a set of independent 2-D projections, a 3-D reconstruction of the object under investigation may be performed by a **3-D Fourier inversion**. As might be gathered, this approach is similar to that used in X-ray diffraction analysis of 3-D protein crystals [c.f. 31], the major difference being that in EM both **amplitudes** and **phases** can directly be recovered from electron micrographs, whereas only amplitudes are contained in X-ray diffraction patterns. It may also be stressed that the Fourier method allows straightforward assessment to be made of the reliability of distinct features and of the achievable resolution in the reconstruction [c.f. 32].

Before combining information from a set of independent 2-D projections collected from a 3-D object, the exact **angular relationships** of the projection images - i.e. the relative orientations of corresponding structural features in the object - have to be established. While a single micrograph may yield a sufficient number of independent views of inherently identical molecules deposited in different orientations on the EM grid to warrant 3-D reconstruction of the molecule to a given resolution, it may, however, not be possible to work out the relative orientations of the different views without going through laborious - and possibly unsuccessful - image analysis steps such as e.g. multivariate statistical data analysis [c.f. 26]. In principle, this problem can be overcome by collecting a **tilt series** (i.e. with known tilt parameters) from a single molecule; in this case, however, the accumulated e⁻ beam damage will prevent us from preserving high resolution structural detail about the molecule. Practically, the situation becomes easiest to handle and warrants the highest resolution to be extracted with a minimal number of tilts, when ordered arrays - possibly with a high degree of symmetry - of the molecule of interest are available. For instance, with a helical array, a single view is usually sufficient to compute a 3-D reconstruction to a given resolution which is defined by the exact helical symmetry of the protein array [c.f. 33]. In the case of 2-D crystalline protein arrays, a set of tilted views (i.e with tilt angles ranging from $0^{o}$ to $\pm 60^{o}$) has to be collected, preferentially each view being recorded from a different array to minimize the effects of radiation damage in the reconstruction [c.f. 34].

## Conclusions and Future Prospects

Despite the instrumental capability of a state-of-the-art EM to resolve single atoms, it is very unlikely that the EM will ever routinely be employed to determine the 3-D structure of protein molecules to near-atomic resolution - to achieve this goal, X-ray diffraction analysis will continue to be the method of choice. This situation will probably remain unchanged, even after we succeed to more systematically obtain highly ordered 2-D protein crystals which, in turn, are a prerequisite for near-atomic resolution **electron crystallography** of proteins to even be attempted. On the other hand, with a comparatively small effort, the EM, in combination with digital image processing, can provide us with 3-D information about the overall size and external shape of protein molecules at the 15-30 Å resolution level using negative staining, in particular with proteins which form ordered arrays (e.g. 2-D crystals or helical polymers). The faithfulness of the thus obtained surface envelope representations of protein molecules strongly depends on the structural detail the negative stain is able to map out. While embedding biological matter in a thin film of amorphous ice is supposedly preserving its 3-D structure in a more native state, it is not clear whether significantly higher resolution structural detail can systematically be gathered from biological specimens than is typically the case when using negative stain [c.f. 35,36] - of course, notwithstanding a few exceptions.

As I see it, the primary use of the EM in structural biology during the next few years will continue to be imaging protein molecules and supramolecular assemblies thereof in the presence of negative stain - and probably increasingly more so in ice - and to apply 2-D and 3-D digital image processing steps to the thus obtained electron micrograph data to yield reproducible molecular detail at the 10-30 Å resolution level. Combining this type of **relative** 2-D/3-D molecular detail with biochemical, biophysical, genetic or immunological information about the specimen under investigation will ultimately reveal important structure-function relationships [c.f. 8,37,38]. Another discipline which will definitely gain momentum in the near future is quantitative STEM such as mass and elemental mapping at the molecular level.

A relatively new and promising instrument which exhibits features that may turn out to be particularly suitable for studying structure-function relationships of biological matter at high resolution is the **scanning tunneling microscope (STM)** recently developed by Rohrer and co-workers at IBM [c.f. 39]. With this instrument a fine metal tip is stepped over the specimen surface by means of piezoelectric ceramic elements, which contract or expand when voltages are applied, thus providing **topographic** or **spectroscopic** information about the specimen at high spatial resolution by measuring **tunnel currents, conductance** or **atomic forces** [c.f.40-44]. Preliminary experiments have demonstrated that not only can the STM be operated in ultra-high vacuum but also at ambient pressure and even in solution [c.f. 45]. The latter possibility may become especially attractive for biological matter, particularly, since the tip may also be employed as a **"nanoelectrode"** to e.g. stimulate or measure ion currents representing distinct functional states of individual membrane channels whose different topographic states, in turn , may be discerned with the tip operating as a **"nanoprobe"** as it is commonly used in STM.

## Acknowledgements

I thank Drs. A. Engel, R. Reichelt and Mr. A. Stemmer for many useful comments and constructive criticism on the manuscript, particularly in conjunction with STEM and STM. The work was supported by grants from the National Institutes of Health (GM-31940 and GM-35171) and by an award from the Maurice. E. Müller-Foundation of Switzerland.

## Literature Cited

1. Beer, M., Frank, J., Hanszen, K., Kellenberger, E. & Williams, R. (1975). Quart Rev. Biophys. **7**, 211-238.
2. Kellenberger, E. & Kistler, J. (1980). In: "Unconventional Electron Microscopy for Molecular Structure Determination", eds. Hoppe W. & Mason, R. (Friedr. Vieweg & Son, Braunschweig/Wiesbaden), 49-79.
3. Baumeister, W. & Vogell, W., eds. (1980). "Electron Microscopy at Molecular Dimensions" (Springer-Verlag, Berlin).
4. Henderson, R. & Unwin, P.N.T. (1975). Nature **267**, 28-32.
5. Jeng, T.-W., Chiu, W., Zemlin, F. & Zeitler, E. (1984). J. Mol. Biol. **175**, 93-97.
6. Henderson, R., Baldwin, J.M., Downing, K., Lepault, J. & Zemlin, F. (1986). Ultra-microscopy **19**, 147-178.
7. Rachel, R., Jakubowski, U., Tietz, H., Hegerl, R. & Baumeister, W. (1986). Ultramicro-scopy **20**, in press.
8. Aebi, U., Fowler, W.E., Buhle, Jr., E.L. & Smith, P.R. (1984). J. Ultrastruct. Res. **88**, 143-176.
9. Engel, A. and Reichelt, R. (1984). J. Ultrastruct. Res. **88**, 105-120.
10. Engel, A., Baumeister, W. & Saxton, W.O. (1982). Proc. Natl. Acad. Sci. USA **79**, 4050-4054.

11. Leapman, R. (1986). J. Electron Microscopy Technique **4**, 95-102.
12. Ottensmeyer, F.P. (1984). J. Ultrastruct. Res. **88**, 121-134.
13. Erickson, H.P. & Klug, A. (1971). Phil. Trans. Roy. Soc. Lond. ser. B, **261**, 105-118.
14. Kistler, J. & Kellenberger, E. (1977). J. Ultrastruct. Res. **59**, 70-75.
15. Taylor, K.A. & Glaeser, R.M. (1976). J. Ultrastruct. Res. **55**, 448-456.
16. Lepault, J., Booy, F.P. & Dubochet, J. (1983). J. Microscopy **129**, 89-102.
17. Taylor, K.A. & Glaeser, R.M. (1975). Science **186**, 1036-1037.
18. Smith, P.R. & Kistler, J. (19777). J. Ultrastruct. Res. **61**, 124-133.
19. Smith, P.R. & Emanuilov Ivanov, I. (1980). J. Ultrastruct. Res. **71**, 25-36.
20. Fowler, W.E. & Aebi, U. (1983). J. Ultrastruct. Res **83**, 319-334.
21. Unwin, P.N.T. & Henderson, R. (1975). J. Mol. Biol. **94**, 425-440.
22. Frank, J. (1973). In: "Advanced Techniques in Biological Electron Microscopy", ed. Koehler, J. (Springer-Verlag, Berlin), 215-274.
23. Frank, J., Goldfarb, W., Eisenberg, D. & Baker, T.S. (1978). Ultramicroscopy **3**, 283-290.
24. Frank, J., Verschoor, A. & Boublik, M. (1981). Science **18**, 1353-1355.
25. van Heel, M. & Frank, J. (1981). Ultramicroscopy **6**, 187-194.
26. Frank, J., Verschoor, A. & Boublik, M. (1982). J. Mol. Biol. **161**, 107-137.
27. Frank, J. (1982). Optik **63**, 67-89.
28. DeRosier, D.J. & Kug, A. (1968). Nature **217**, 130-134.
29. Crowther, R.A., DeRosier, D.J. & Klug, A. (1970). Proc. Roy. Soc. Lond. ser. A **317**, 319-340.
30. Herman, G.T. (1979). In: "Topics in Applied Physics" (Springer-Verlag, Berlin), Vol. **32**.
31. Blundell, T.L. & Johnson, L.N. (1976). Protein Crystallography (Academic Press, New York).
32. Klug, A. & Crowther, R.A. (1972). Nature **238**, 435-440.
33. Smith, P.R. Aebi, U., Josephs, R. & Kessel, M. (1976). J. Mol. Biol. **106**, 243-275.
34. Smith, P.R., Fowler, W.E., Pollard, T.D. & Aebi, U. (1983). J. Mol. Biol. **167**, 641-660.
35. Beer, M., ed. (1984). Symposium on "Electron Crystallography of Macromolecules" in: Ultramicroscopy **13**, Nos. 1/2.
36. Stewart, M. & Vigers, G. (1986). Nature **319**, 631-636.
37. Unwin, P.N.T. & Ennis, P.D. (1984). Nature **307**, 609-613.
38. Kistler, J., Aebi, U., Onorato, L., ten Heggeler, B. & Showe, M.K. (1978). J. Mol. Biol. **126**, 571-589.
39. Binnig, G. & Rohrer, H. (1982). Helv. Phys. Acta **55**, 726-735.
40. Binnig, G. & Rohrer, H. (1985). Sci. American **235**(2), 40-46.
41. Robinson, A.L. (1985). Science **229**, 1074-1076.
42. Salvan, F., Fuchs, H., Baratoff, A. & Binnig, G. (1985). Surface Sci. **162**, 634-639.
43. Hamers, R.J., Tromp, R.M. & Demuth, J.E. (1986). Phys. Rev. Lett. **56**, 1972-1975.
44. Quate, C.F. (1986). Physics Today, August 86, 26-33.
45. Sonnenfeld, R. & Hansma, P.K. (1986). Science **232**, 211-213.

# MAGNETIC RESONANCE IMAGING IN MEDICINE

P. Boesiger
Institute of Biomedical Engineering of the University of Zurich and
Swiss Federal Institute of Techology
Moussonstrasse 18, CH-8044 Zurich, Switzerland

## Introduction

Since their discovery and their initial applications by Bloch and
Purcell in 1946 the phenomena of magnetic resonance have been widely
used by chemists, biochemists and physicists in the study of both in-
organic and organic compounds. Over the last 25 years with the increasing interest in the fundamental chemistry and physics of biological processes, magnetic resonance spectroscopy has also been applied
in the study of organized biological systems. Its use has yielded considerable amounts of information about the life processes both in
simple organisms and in isolated organs of more complex living systems. Since the early seventies magnetic resonance phenomena have
been used to obtain non-invasively information about metabolic processes in animals and in human beings.

Two of the most spectacular developments during the last fifteen years
based on magnetic resonance are the technique of nuclear magnetic resonance imaging and the localized magnetic resonance in-vivo spectroscopy. Nuclear magnetic resonance imaging allows to produce two-dimensional slice images of the human body in any arbitrary direction or
even to acquire three-dimensional image data. In contrast to X-ray tomography or ultrasound echography, where absorption coefficients of X-
rays or scattering properties of ultrasound waves are visualized, the
magnetic resonance imaging delivers more dimensional information. According to the sequence of radiofrequency pulses used for the excitation of the commonly used hydrogen nuclei, the density of the nuclei
or their relaxation times $T_1$ or $T_2$ or any combination of these three
parameters are shown. Special techniques also allow to visualize or
even to quantify blood flow in greater vessels.

The main advantages of magnetic resonance imaging in comparison to conventional imaging techniques as X-ray tomography or ultrasound echography are:
- improved soft tissue contrast
- no bone artefacts
- arbitrary direction of slices
- no known hazard
- visualization and quantification of blood flow
- biochemical information

The disadvantages are:
- no visualization of compact bone
- high costs
- large space requirements
- low experience

At least in part these disadvantages become less and less important because the experience for the interpretation of the images is continuously growing and the space requirements become smaller and smaller with the development of shielding equipment for the magnetic stray field.

Magnetic resonance phenomena

Nuclear magnetic resonance is based on the fact that a series of nuclei show a spin and in coupling with that spin a magnetic dipolar moment (fig. 1a). If the nuclei are placed in an external magnetic field - we restrict to nuclei with spin 1/2 for the moment - their magnetic moments will align parallel or antiparallel to the external field under continuous precession around the field direction with their characteristic Larmor frequency $\omega_0$, which can be calculated to be

$$\omega_0 = \gamma B_0$$

$B_0$ represents the magnetic field strength. $\gamma$ is a constant of proportionality which is called the gyromagnetic ratio. Its value is dependent of the type of nucleus involved.

The ratio of spins parallel to spins antiparallel to the external field is given by Boltzman's law. For field strengths as they are usually applied for magnetic resonance imaging (0.2 ... 1.5 Tesla) the

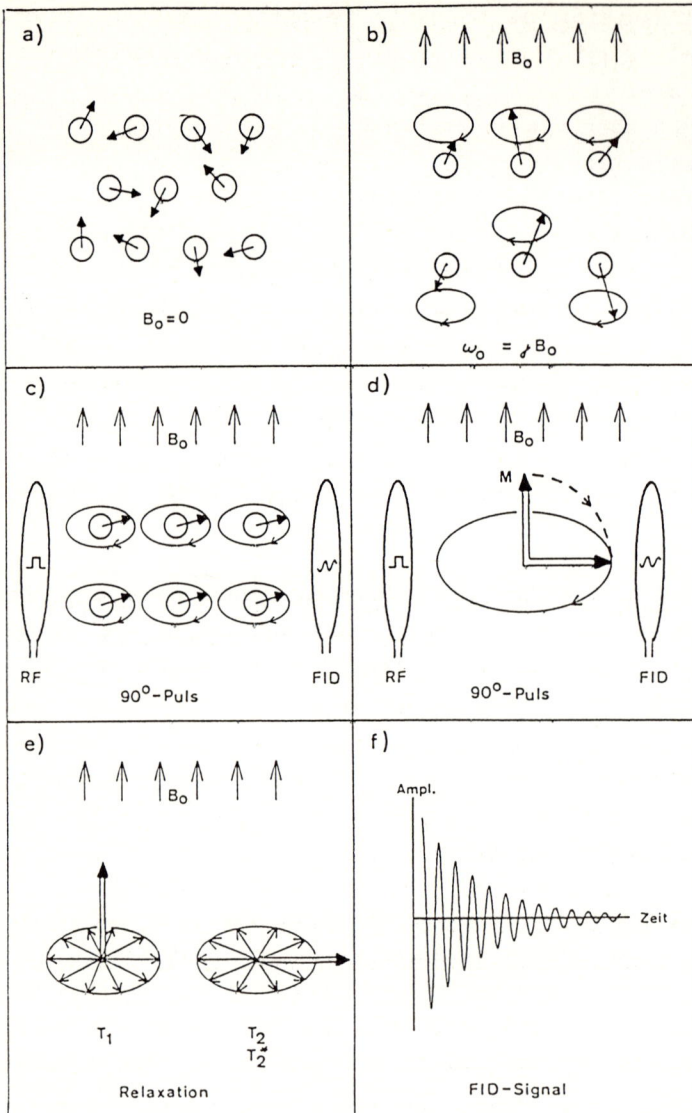

Fig. 1: Fundamentals of a NMR experiment
  (a)  random orientation of the nuclear dipolar moments
  (b)  alignment with an applied magnetic field
  (c)  and (d) excitation of the dipolar moments and the macroscopic magnetization vector
  (e)  relaxation processes
  (f)  free induction decay signal induced by the precession and the relaxation of the magnetization vector

energy difference between the two states is very small. Therefore the difference of the two populations normalized with the number of nuclei is only about five parts per million leading to a poor signal-to-noise ratio in the experiments. This poor signal-to-noise ratio is an inherent problem of NMR and sets certain limits to the applications of the technique.

Since measurements cannot be made on single nuclei we rather look at the macroscopic magnetization of the entire probe or for magnetic resonance imaging of an individual volume element of the probe. While a single nucleus has to be treated according to the laws of quantum mechanics the behaviour of a population of spins and with it the macroscopic magnetization vector can be described by the laws of classical physics. In NMR experiments it is the precession of this magnetization vector that is observed (fig. 1d).

By the application of a radiofrequency pulse in the plane perpendicular to the static magnetic field $B_0$, whose frequency corresponds to the Larmor frequency of the nuclei, the magnetization is flipped under continuous precession around the z-axis from its steady state by a certain tip angle. By a carefully selected pulse length and pulse amplitude the tip angle may be 90 degrees; the corresponding pulse is called a 90 degree pulse. In the same way it is possible to apply pulses producing any arbitrary pulse angle. After a 90 degree pulse the magnetization is perpendicular to the static field; it experiences a torque producing a precession of the magnetization in the plane perpendicular to the field. It induces a sinusoidal signal in a receiver coil (fig. 1f), whose initial amplitude is proportional to the magnetization and therefore, if the system is initially in its thermodynamical equilibrium, to the number of the spins within the probe or the volume element. The signal induced in the coil, the so called "free induction decay signal", disappears by reason of two relaxation mechanisms (fig. 1e): The spin spin interaction causes a stochastic dephasing of the magnetic moments precessing all with the same phase after the 90 degree pulse (fig. 1c) with the so called spin spin relaxation time $T_2$ of 50.. 100 ms typically for biologic samples. In conventional magnetic resonance imaging experiments the dephasing is much faster because of the inhomogeneities of the magnetic field. They are mainly produced by magnetic field gradients which are applied to achieve spatial resolution of the NMR signals. But these systematic dephasing effects can be overcome applying a 180 degree pulse to produce a refocussing of the spins, a so called spin echo. In addition to the spin

spin interaction the spin lattice interaction provides for an energy transfer from the spin system to its surrounding, the so called lattice, resulting in a realignment of the magnetization in the field direction with a characteristic time constant $T_1$ of 500..1200 ms typically (fig. 1e).

This behaviour of the magnetization under the influence of static and time dependant magnetic fields and the relaxation processes can be described phenomenologically by the classical Bloch equations:

$$\frac{d\vec{M}}{dt} = \gamma \, (\vec{M} \times \vec{B}) + Rel.$$

With the static field $B_0 = (0,0,B_z)$ and the time dependant field $B_1(t) = (B_x(t), B_y(t), 0)$ this leads to

$$\frac{dM_x}{dt} = \gamma \, (M_y B_z - M_z B_y(t)) - \frac{1}{T_2} M_x \qquad .$$

$$\frac{dM_y}{dt} = \gamma \, (M_z B_x(t) - M_x B_z) - \frac{1}{T_2} M_y$$

$$\frac{dM_z}{dt} = \gamma \, (M_x B_y(t) - M_y B_x(t)) - \frac{1}{T_1} (M_z - M_0)$$

Neglecting the relaxation terms and for short radiofrequency pulses in x-direction

$$B_1(t) = (B_1 \cos \omega t, \, 0, \, 0)$$

the equations can be solved by a transformation in a coordinate frame rotating around the z-axis.

Magnetic resonance imaging

The basic principle of MR imaging is based on the application of well defined linear magnetic field gradients, so that the strength of the magnetic field becomes a linear function of one coordinate, e.g. with a field gradient $G_z$ of the z-coordinate. Therefore the Larmor frequency which is proportional to the strength of the magnetic field becomes a function of the z-coordinate too:

$$\omega_0 = \gamma \, B_z = \gamma \, (B_0 + z \, G_z)$$

Based on this fact many different procedures can be derived for imaging of 2d slices of a 3d object or for acquisition of full 3d image

data. The method mainly used today is the 2d Fourier method, which basically was proposed by Ernst at ETH Zurich:

In a first step selective excitation is applied for the excitation of
the spins of a selected slice. The selective excitation method employs
a field gradient, e.g. in the z-direction. As a consequence the resonance frequency of the nuclei increases along this axis. The slice to
be selected thus contains a certain narrow band of Larmor frequencies.
The excitation of the spins with electromagnetic radiation of the same
narrow band frequencies by means of a specially shaped RF pulse con-
fines the excitation to that particular slice. The slice thickness depends on the bandwidth of the excitation pulse and on the magnitude of
the gradient. To produce sagittal or coronal images the gradient has
to be applied in the x- or y-direction respectively.

In a second step after selection of the slice the NMR response signal
has to be labelled spatially in order to obtain spatial resolution
within the slice. It is achieved  by the subsequent application of a
so called phase encoding gradient $G_y$ and a frequency encoding or
readout gradient $G_x$ (fig. 2).

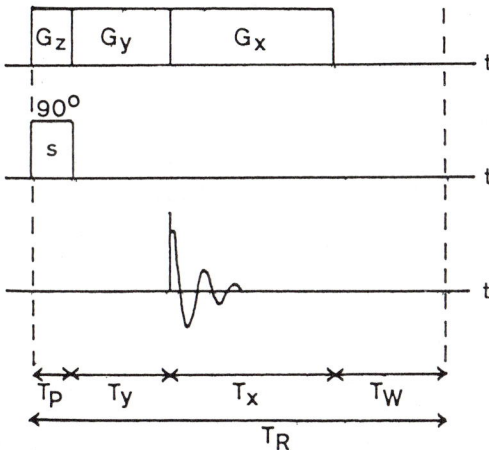

Fig. 2: Fundamental radiofrequency and gradient pulse sequence for two
        dimensional Fourier imaging

After the application of the slice selection gradient $G_z$ and the corresponding RF excitation pulse, all the spins within the slice precess
in phase with their Larmor frequency. During the period of the phase
encoding gradient nuclei will have modified their Larmor frequency according to their position in the y-direction. When the gradient is

switched off, the spins show different precession phases according to their y-direction. If then the readout gradient $G_x$ is applied, the frequency of precession varies according to the x-coordinate of the nuclei. The signal induced in the receiver coil during the application of $G_x$ is the composite of the signals from all the nuclei precessing at different phases according to their y-coordinate and at different frequencies according to their x-coordinate.

If we write down that analytically, the signal of one volume element $(i,j)$ with the coordinates $(x_i,y_j)$ at the sampling time intervals $T_{xk}$ after the application of a phase encoding gradient $G_{yl}$ and after demodulation with the Larmor frequency $\omega_o = \gamma B_o$ becomes (see also fig. 2)

$$\sigma_{k\ell}^{ij} \doteq \sigma(x_i, y_j, T_{xk}, G_{y\ell})$$

$$\sim \tilde{g}(x_i, y_j, z_o) \exp\left\{\bar{i}\left(\underbrace{\gamma G_x T_{xk} x_i}_{k_{xk}} + \underbrace{\gamma G_{y\ell} T_y y_j}_{k_{y\ell}}\right)\right\}$$

The signal of the entire slice will be

$$s_{k\ell} \doteq s(k_{xk}, k_{y\ell})$$

$$\sim \sum_{j}^{N}\sum_{i}^{N} \tilde{g}(x_i, y_j, z_o) \exp\left\{\bar{i}\left(k_{xk} x_i + k_{y\ell} y_j\right)\right\}$$

If the signal is sampled in $k=1..N$ chanels and the experiment is repeated for $l=1..N$ profiles with different phase encoding gradients the image can be reconstructed by a 2d Fourier transformation:

$$\tilde{g}(x_i, y_j, z_o) \sim \sum_{k}^{N}\sum_{\ell}^{N} s_{k\ell} \exp\left\{-\bar{i}\left(k_{xk} x_i + k_{y\ell} y_j\right)\right\}$$

$$= \sum_{\ell}^{N}\left[\sum_{k}^{N} s_{k\ell} \exp\left\{-\bar{i} k_{xk} x_i\right\}\right] \exp\left\{-i k_{y\ell} y_j\right\}$$

By cyclic exchange of the three gradients $G_x$, $G_y$, $G_z$ the slice direction can be altered. If instead of the selective excitation procedure a second phase encoding gradient in z-direction is applied, 3d image data can be acquired which is reconstructed by 3d Fourier transformation.

In practice it is not possible to acquire data of high resolution images with the aid of the pulse and gradient sequences discussed above. Because of the field inhomogeneities produced by the gradients the spins dephase very fast, so that the signal disappears. Therefore spin echo signals are induced by an additional 180 degree pulse and an additional gradient in the x-direction.


Image contrast and tissue characterization

The signal intensity is determined at constant field and constant sample temperature by the number of spins within the probe or the volume element, if we assume thermal equilibrium at the time before the excitation is performed. For imaging purposes the experiment has to be repeated e.g. 256 times for a resolution of 256 picture elements in both directions. To achieve thermal equilibrium between two subsequent excitiations one has to wait for a time interval of about five times $T_1$ of the tissue with the longest relaxation time, that means of about 6 sec, between two excitations. Thus the time for the acquisation of one image would be about 25 min. In addition these images show poor contrast because the grey levels are determined by the number of spins of each voxel which is quite similar for different types of tissue. If the interval is shortened partial saturation occurs; the signal amplitude becomes a function of the spin density, the relaxation time $T_1$ and the repetition time $T_R$ and will be given by

$$\sigma \ \sim \ \tilde{s} \ [ \ 1 - exp \ (- T_R \ / T_1 \ ]$$

If in addition we take into consideration the usually applied spin-echo detection, the signal amplitude also becomes dependant on the echo time $T_E$ and the relaxation time $T_2$ respectively

$$\sigma_{SE} \ \sim \ \tilde{s} \ [ 1 - exp \ (- T_R / T_1 )] \ exp \ (- T_E / T_2 )$$

Another pulse sequence often applied is the inversion recovery sequence. It is defined by an initial 180 degree pulse which inverts the magnetization, followed by a 90 degree excitation pulse after the inversion time interval $T_I$. The signal amplitude can be calculated to be

$$\sigma \ \sim \ \tilde{s} \ [ 1 - 2 \, exp \, (- T_I / T_1 ) + exp \, (- T_R / T_1 )]$$

for repetitive excitation or

$$\sigma_{IR} \sim \tilde{s} \left[ 1 - 2 \exp(-T_I/T_1) + \exp(-T_R/T_1) \right] \exp(-T_E/T_2)$$

if spin echo detection is applied.

As it can easily be seen from these formulas the signal amplitude and therefore the relative gray scale level of the tissue on the image depends in a complicate manner on the tissue parameters rho, $T_1$ and $T_2$, from the pulse sequence used for the excitation and its timing parameters $T_R$, $T_E$ and $T_I$. As a consequence the contrast of different type of tissue also depends on these timing parameters and can be modified over a wide range by the selection of the parameter values.

To achieve quantitative tissue characterization it is desirable to acquire images visualizing the tissue parameters rho, $T_1$ and $T_2$ separately rather than complicated combinations of these parameters and the pulse sequence parameters. Images visualizing only one of these tissue parameters can be calculated from appropriate sequences of images. A $T_2$ image is obtained by a pixel by pixel fitting procedure from a set of spin-echo images with various echo times. Such a set may be acquired e.g. by a train of subsequent 180 degree pulses within time intervals of $T_E$. In a similar way $T_1$ images can be fitted from a set of inversion recovery images with different inversion times $T_I$. Finally extrapolation of the $T_1$ recovery curve to full recovery and to echo times $T_E = 0$ yields spin density images.

## Fast imaging sequences

The time span for the acquisition of an NMR image is determined by the desired spatial resolution and the repetition time which cannot be made arbitrary short because of saturation effects. Therefore that part of the repetition time which consists of the recovery period for the spin system, is virtually a waiting time. During this waiting time it is possible to collect image profiles of other slices of the probe. By this multislice technique images can be acquired for a number of slices within a time interval similar to that required for a single plane, e.g. for 8 or 16 slices within 2...8 min.

Another possibility to avoid saturation is to use flip angles smaller than 90 degrees, e.g. 30 degrees for excitation of the spins. Thereby always a magnetization component in the direction of the static field is maintained and the repetition time can be reduced down to e.g. 30 ms. If additionally the echo pulse is replaced by a gradient inversion it becomes possible to acquire fully resolved images within a few seconds.

Chemical shift and spectroscopy

Up to now we assumed that besides of some stochastic dephasing effects identical nuclei show identical Larmor frequencies. Indeed they all have the same gyromagnetic ratio, leading to identical Larmor frequencies if they feel all the same magnetic field strength. But the electrons encircling the atomic nucleus shield the external magnetic field to a certain extent, the nucleus being subjected to a slightly weaker field than that applied. Therefore the Larmor frequency depends on the configuration of the electron cloud, i.e. it depends on the molecular structure. Thus identical nuclei being part of different kinds of molecules or occupying different places in a molecule emit free induction decay signals with slighly different frequencies. Consequently the NMR spectrum can show various resonance lines. The frequency shifts to a well defined reference line expressed as parts per million are called "chemical shifts". The peak areas in a spectrum are proportional to the number of nuclei contributing·to the signal.

In conventional NMR spectroscopy the probe under investigation is placed in a strong and homogeneous magnetic field. The signal of the entire probe is picked up by a receiver coil. In whole body in-vivo spectroscopy the problem arises of how to constrict the volume that contributes to the signal. It doesn't help anything to acquire e.g. a phosphorus signal of an entire human being. One rather wants to acquire a localized spectrum of an organ such as the brain, the liver, the kidney, the heart muscle or even only of a certain area of an organ. Often this problem is solved by the use of surface coils. As a result of their geometry, these coils have a limited sensitivity range. The sensitivity range can further depend on the timing within the pulse and gradient sequences applied. Another possibility is to reduce the volume by selective excitation techniques, as it was shown by Aue and Seelig, den Hollander and Luyten, and other authors.

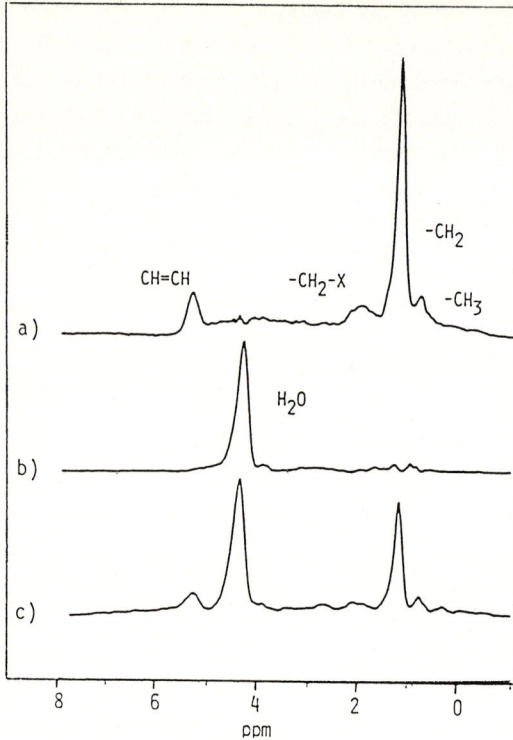

Fig. 3: $^1$H spectra of a human calf acquired by selective excitation
techniques on a whole body system
(a)   spectrum of bone marrow
(b)   spectrum of muscle tissue
(c)   spectrum of muscle tissue and subcutaneous fat

The techniques are still under development. Fig. 3 shows several
$^1$H spectra of a human calf acquired with the selection excitation
technique called SPARS on a whole body system. The goal is a combination of magnetic resonance imaging with localized spectroscopy to
identify very precisely the volume under spectroscopic investigation.
For the time beeing the localized spectroscopy for human applications
is mainly restricted to $^1$H and $^{31}$P by sensitivity reasons; in animals
also $^{13}$C spectra are in evaluation.

Reference:

Bösiger, P.: Kernspin-Tomographie für die medizinische Diagnostik.
          B.G. Teubner, Stuttgart, 1985. ISBN 3-519-03066-7

# DIGITAL IMAGE PROCESSING

Murat KUNT

Signal Processing Laboratory
Swiss Federal Institute of Technology, Lausanne
16 Ch. de Bellerive
CH - 1007 Lausanne, Switzerland

Two decades ago, it would be reasonable to write a small chapter under this title. Today, the field of digital image processing is so large that there are already several books covering differents aspects of it such as [1]-[12] to name a few. Each year, several international conferences attracting increasingly large number of participants are held [13]-[15]. The explosion of the field in two decades show its importance. In such a case what one can write in a small chapter ? The text will necessarily be biased by the authors interests. To be as objective as possible it is preferable to give a summary and indicate relevant references.

Images, along with sounds, are our main carriers for communication with each other, and, more contemporarily, with machines. The enormous impact of photography and later on that of the broadcast TV in our societies are obvious. The question one may ask is then the following : what digital image processing does that the conventional processings (classical photography and conventional TV) does not ? The answer is so long that one should again refer to [1]-[6]. We shall try to summarize here a possible and perhaps biased answer. The most condenced answer is the title : digital image processing. We all know what images are. Explainations are needed for the first word and the last word.

If a signal, carrying information, evolves continuously in time, it is called analog signal. Its values can be determined any time. If such a signal does not vary more rapidly than a given limit, it is not necessary to handle all its values. Samples taken at a specified pace are sufficient to convey the same information, since the analog signal can be recovered from its samples exactly, whenever needed. Each sample may have any value within a finite range. Very often, such a infinite precision on the magnitude of each sample is not necessary, simply because we cannot even measure it. A finite number values can then be assigned to these samples. To increase the precision, one should increase the number of these preassigned values. A signal described by a set of samples whose magnitudes are quantized this way is called digital signal. Analog signals must be processed by analog systems and digital signals must be processed by digital systems, the same way one cannot use a hammer to fix a screw and a screw driver to fix a nail. Analog processing systems, such as chemicals to develop a film, electronic amplifiers or modulators to transmit a video signal, are very limited in their processing possibilities. For example, by changing the exposure time or the proportion of some chemicals or by using special masks, one can change the contrast, the granularity or the appearence of a picture in conventional photography. But, it impossible, with the same techniques, for example, to detect automatically the presence of a car on the picture, or to combat the effects of a bad focusing.

In contrast with analog techniques, digital systems offer unlimited processing possibilities. Early digital processing systems were big general purpose computers. They are very versatile. Any processing that can be translated into a computer program can be implemented. Human imagination is so fertile that it was not difficult to design complexe digital processings to saturate these machines with several hours of computation time. It was also very easy to find a large volume of data to be processed. These constraints created a chain reaction. Progress in digital electronic technology offer faster machines, more complexe processings are implemented and designed requiring in turn even faster machines. A good balance is also established between programming flexibility at necessarily lower speeds and hardwiring rigidity at higher speeds. Specialized computer architectures are beeing developped, going away from classical Von Neumann architecture of general purpose computers, to speed up processing still maintaining and improving flexibility.

The principle to obtain a digital image is the following. A three dimensional scene is projected onto a two dimensional plane. The image on the plane is subdivided into a set of very small areas tessalating the entire plane. These areas can be for example squares, circles or any reasonable simple geometric shape. The amount of light transmitted or reflected over each individual area is measured by integration. The numerical values obtained are then quantified one by one with a finite set of preassigned values, leading thus to a digital image. In other words, analog light data are sampled in space and quantized in brightness. For example, the conventional TV systems have already the sampling in the vertical dimension of the image. They produce and analog signal by scanning the scene on a line-by-line basis. Such a signal can be sampled in the other dimension and quantized. The new so-called CCD cameras give directly a digital image. These images are stored temporarily on a compatible carrier (digital electronic memory, computer disk, magnetic tape, etc.) to be processed. The choice of the sampling size and the number of preassigned quantization values is of primary importance. Although the sampling size may be related to theoretical studies, in practice often it is chosen "small enough, to be invisible ". One may see the dot pattern of a digital image if it is observed on a short distance. The same image will look as an ordinary analog image if the observation distance is increased. Human visual system performs a very useful digital to analog conversion. The number of preassigned quantization values are determined on the basis of psychovisual experiments and simple mathematical expressions.

The word 'processing' is semantically very reach. For this reason, classifications are established to characterise various possible processings. For example, we may distinguish a processing which transforms the input image into another output image from another processing which acts on the input image to extract a numerical value or a decision.

Another distinction can be made on a more mathematical basis between linear and non linear processing methods. The well known superposition principle is the appropriate criterion. Linear processing methods are widely used because they are easy to model, to analyse and to implement. Their scope remains, however, rather limited. For example a linear processing ban easily be implemented to remove slowly varying (spatial low frequency) components in order to emphasize abrupt changes in brightness. Another example is that of measuring the similarity between two images using correlation techniques. The frequency content of an image can be analysed using the Fourier transform as still another example. These thechniques are now well established, rigorously analysed and efficiently implemented.

To enlarge the scope and the possibilities, non linear methods are in order. The main difficulty with them is the weakness of the mathematical rigor requiring intensive experimental work to determine their validity, robustness and range. A typical example is contour extraction. Contours are defined as abrupt changes in brightness. As mentioned earlier, they can, in principle, be detected by linear high pass filtering. But, after filtering, a decison should be made on each picture element (sample, or pixel) labeling it as contour or non contour point, a non linear operation. Even with a very sophisticated linear filter, not only contours but also a lot of small artefacts are also detected. Better performing contour exctractors are nonlinear but their performances are measured often subjectively and always uncompletely. A contour exctractor may give very good results on one image and perform poorly on another. For this reason, there are at least ten different contour extraction methods.

More complexe methods, thus more difficult to analyse and interpret, are used for human-like tasks. Our ability to regognize a person or an object at very high speeds is so normal that we do not even pay attention to its mechanism. If the same recognition has to be done by a machine, the required processing becomes very complexe and time consuming. When digital imege processing is used for recognition and inspection, one enters into fields called pattern recognition and scene analysis. In the author's opinion, that is where the most important research effort is spend today.

## References .

[1] H.C. Andrews, "Computer Techniques in Image Processing", Academic Press, N.Y., 1970.

[2] A. Rosenfeld, "Picture Processing by Computer", Academic Press, N.Y. 1969

[3] W.K. Pratt, "Digital Image Procesing", John Wiley and Sons, N.Y. 1978.

[4] A. Rosenfeld et al., "Digital Picture Analysis", Springer Verlag, Berlin, 1976.

[5] H.C. Andrews and B.R. Hunt, "Digital Image Restoration", Princtice Hall, N.J., 1977.

[6] R.O. Duda and P.E. Hart, "Pattern Classification and Scene Analysis", John Wiley and Sons, N.Y. 1973.

[7] R.C. Gonzales and P. Wintz, "Digital Image Processing", Addison Wesley Punlishing Co., 1977

[8] K.R. Castelman, "Digital Image Processing", Princtice Hall, N.J. 1979

[9] T.S. Huang, (Edt), "Image Sequence Analysis", Springer Verlag, Berlin 1981

[10] J.C. Simon and B.R. Haralick, "Digital Image Processing", D. Reidel Publishing Co., 1981

[11] D.H. Ballard and C.M. Brown, "Computer Vision", Printice Hall, N.J. 1982

[12] A. Rosenfeld, (Edt.), "Image Modeling", Academic Press, 1981

[13] IEEE Computer Society, Pattern Recognition and Image Processing Conference.

[14] International Joint Conference on Patter Recognition.

[15] International Picture Coding Symposium.

# PHYSICS OF BIOLOGICAL MEMBRANES

Ole G. Mouritsen

Department of Structural Properties of Materials
The Technical University of Denmark
Building 307, DK-2800 Lyngby
Denmark

## Abstract

The biological membrane is a complex system consisting of an aqueous biomolecular planar aggregate of predominantly lipid and protein molecules. At physiological temperatures, the membrane may be considered a thin ($\sim 50 \text{Å}$) slab of anisotropic fluid characterized by a high lateral mobility of the various molecular components. A substantial fraction of biological activity takes place in association with membranes. As a very lively piece of condensed matter, the biological membrane is a challenging research topic for both the experimental and theoretical physicists who are facing a number of fundamental physical problems including molecular self-organization, macromolecular structure and dynamics, inter-macromolecular interactions, structure-function relationships, transport of energy and matter, and interfacial forces. This paper will present a brief review of recent theoretical and experimental progress on such problems, with special emphasis on lipid bilayer structure and dynamics, lipid phase transitions, lipid-protein and lipid-cholesterol interactions, intermembrane forces, and the physical constraints imposed on biomembrane function and evolution. The paper advocates the dual point of view that there are a number of interesting physics problems in membranology and, at the same time, that the physical properties of biomembranes are important regulators of membrane function.

## Table of Contents

## 1.   INTRODUCTION

### 1.1   Biological membranes : What they are

The biological membrane is the most common cellular structure in living matter.[1] Every living cell is bounded by a plasma membrane and higher cells have in addition a number of internal membranes which divide the cell into specialized compartments. A schematic drawing of a cross section of a biomembrane is shown in Fig. 1.   This

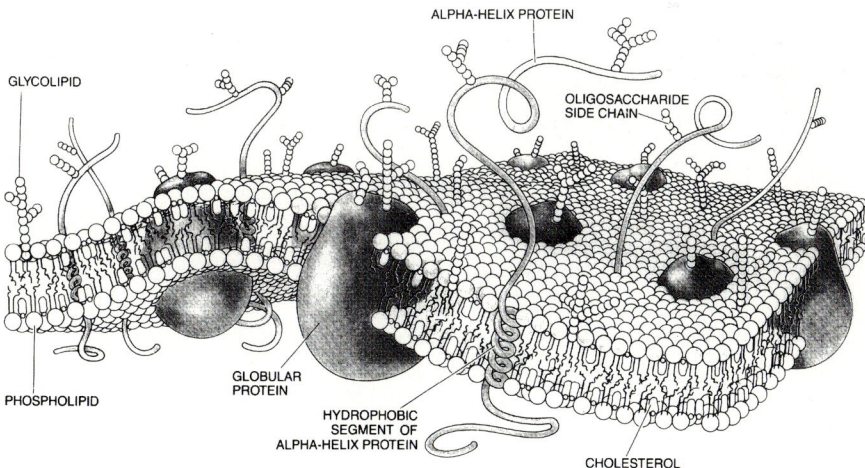

Fig. 1.   The fluid-mosaic model of a biological membrane.   [Reproduced with permission from *Scientific American*, cf. Ref. 3.]

figure reflects the conventional wisdom of the so-called Singer-Nicolsen model[2] of biomembranes which describes the membrane as a fluid-mosaic pseudo-two-dimensional aggregate of a lipid bilayer with embedded membrane-bound proteins.

Lipid molecules are amphiphiles, that is they have mixed feelings about water. In the one end they consist of one or two hydrophobic hydrocarbon chains and in the other end they have a hydrophilic polar head. Different lipid species differ with respect to hydrocarbon chain length and degree of saturation, as well as nature of the polar head. Natural membranes typically contain a large number of different lipid species,[4,5] a phenomenon referred to as lipid diversity. The lipid molecules have many properties in common with soaps and detergents and they spontaneously form macroscopic aggregates when mixed with water.[6] It is a remarkable fact that such aggregates do not involve covalent binding between the molecules – the aggregate owes its existence solely to the hydrophobic effect. The structural formulas for two particular lipid species which will be referred to frequently in the following are shown in Fig. 2 :

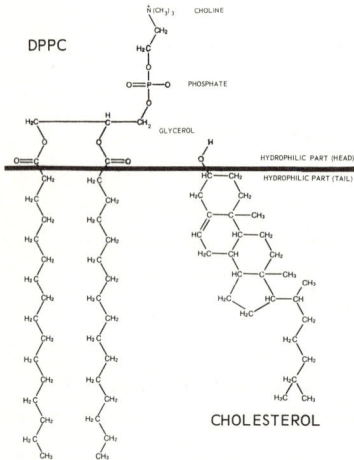

Fig. 2. Lipid molecules of the biological membrane: Dipalmitoylphosphatidylcholine (DPPC) and cholesterol. The lipid molecules are amphiphilic with a water soluable polar head and a hydrophobic tail.

dipalmitoylphosphatidylcholine (DPPC) and cholesterol. In the case of DPPC, the polar head is very bulky and the two saturated palmitic acid chains, each with 15 C-C bonds, are fully flexible. Conversely, the polar head of cholesterol is just the hydroxyl group and the hydrophobic tail is predominantly a rigid sterol skeleton. Obviously these two lipids will serve very different purposes in the membrane. A phospholipid related to DPPC is DMPC which has only 13 C-C bonds in the chains.

## 1.2 Biological membranes : What they do

The membrane serves both a number of passive and active functions. Firstly, it provides compartmentalization of the living matter and separates inside from outside by a barrier with low permeability to ions and large molecules. Secondly, it acts as a suitable anchor place for the 'machines' of the cell membranes, i.e. the proteins, ion channels, and enzymes.

However, the membrane not only acts as a passive container, it also supports a substantial fraction of biological activity, e.g. transport of matter, energy trans-duction, growth, intercellular recognition, immunological response, nerve processes, and biosynthesis.

## 1.3   The role of the physicist in membranology

Despite the fundamental importance of biological membranes to many life processes, the quantitative physical knowledge about membranes has been rather limited compared to the information which is presently available about other cellular components.  However, in recent years an increasing number of physicists have taken a growing interest in the study of biomembranes using experimental and theoretical methods well-known from condensed matter physics.  The approach of the physicists may be considered as being complementary to the more conventional, phenomenological biochemical and phys-iological study of the functioning of membranes.  Both approaches, however, endeav-our to characterize the relationship between the physiological *function* and the phys-ical and molecular *structure* of biological membranes.

In this review, we shall focus on recent progress which has been made in pro-viding information on the physical properties of biomembranes in relation to lipid bi-layer thermodynamics, structure, and dynamics; lipid phase transitions and phase equilibria; effects due to interactions with proteins and cholesterol; and intermembrane forces.

Within this sphere, a major contribution of physicists to biomembrane science has been a quantitative characterization of *general* properties of model membranes, such as few-component synthetic lipid membranes and reconstituted cell membranes with a single "impurity", e.g. proteins, alcohols, cholesterol, drugs, or antibiotics.  The character-ization of lipid membrane thermodynamics, structure and molecular dynamics has in-volved a great variety of experimental techniques ranging from classical calorimetry,[7] micromechanics,[8] x-ray[9] and neutron scattering,[10] to spectroscopic techniques such as electron spin resonance,[11] nuclear magnetic resonance,[11,12] fluorescence polariza-tion,[13] and Raman spectroscopy.[14]  The elucidation of the structure of integral mem-brane proteins[15] has proceeded by means of image analysis of electron micrographs[16] or by x-ray diffraction.[17]

The progress on the experimental side has highlighted a need for a parallel the-oretical activity.  This need has been met by a number of attempts to theoretically model e.g. lipid membrane phase transitions and phase equilibria,[18-22] interactions with protein and cholesterol,[23,24] and intermembrane interactions.[25,26]  The rather advanced stage of the experimental programmes has in many cases called for a rather detailed and complicated modelling, in particular of lipid membrane phase transitions and hydrocarbon chain dynamics.

## 2. LIPID MEMBRANE STRUCTURE AND DYNAMICS

### 2.1 Lipid membranes in space and time

When immersed in water, lipid molecules form macroscopic aggregates which, in the case of excess water, frequently appear as bilayers, either in a multilameller dispersion or as closed-membrane vesicles, Fig. 3. Such bilayers constitute the simplest

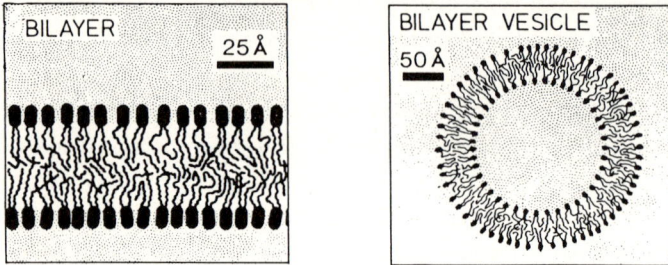

Fig. 3. Schematic drawings of lipid bilayer aggregates: planer lipid bilayer and bilayer vesicle [After Ref. 27].

class of models for biological membranes.

Depending on the lipid chain length,[28] the bilayer thicknesses lie in the range from 50-100Å with a polar-head group region of about 5Å. The width of the hydration layer is of the order of 10Å, depending on the polar head group type. Vesicle sizes range to a diameter of up to about $10^4$Å. This number should be compared to the linear extensions of real cells which typically lie in the range $10^5$–$10^6$Å, with procaryotic cells in the low end and encaryotic cells in high end.

At physiological temperatures the lipid membranes are fluid systems, that is the membrane components have high lateral mobility (see, however, Sec. 2.4). It is very important to note that as fluid systems, membranes are highly anisotropic with a diffusion anisotropy of the order of $10^7$! This is seminal for maintaining membrane asymmetry. Referring to the standard expression for lateral diffusion

$$\langle r^2 \rangle = 4D_\ell t \ ,$$ (1)

typical lipid and protein diffusion constants are found to be $D_\ell \sim 10^{-8} cm^2/sec$ and $10^{-10} cm^2/sec$, respectively. For an average cell this implies that a lipid molecule can travel around the cell in 20 sec, whereas the same excursion for an integral membrane-bound protein may take 20 min. The corresponding rotational diffusion characteristics, cf.

$$\langle \theta^2 \rangle = 2D_r t$$ (2)

is found to be $D_r \sim 10^8 sec^{-1}$ for lipids and $D_r \sim 10^4 sec^{-1}$ for proteins. Thus, a lipid may rotate once around its long axis while it travels over its own diameter, whereas a protein only performs one rotation while diffusing over a distance corresponding to ten protein diameters. Hence, the biological membrane is an extremely lively system!

## 2.2   Lipid bilayers are model membranes with phase transitions

A selection of by now almost classical experimental results on the thermal behaviour of wet lipid bilayers is presented in Fig. 4 for bilayers of DPPC and DMPC. All experiments indicate that the bilayers display striking thermal anomalies. Fig. 4a shows that the specific heat exhibits a pronounced peak at the same temperature

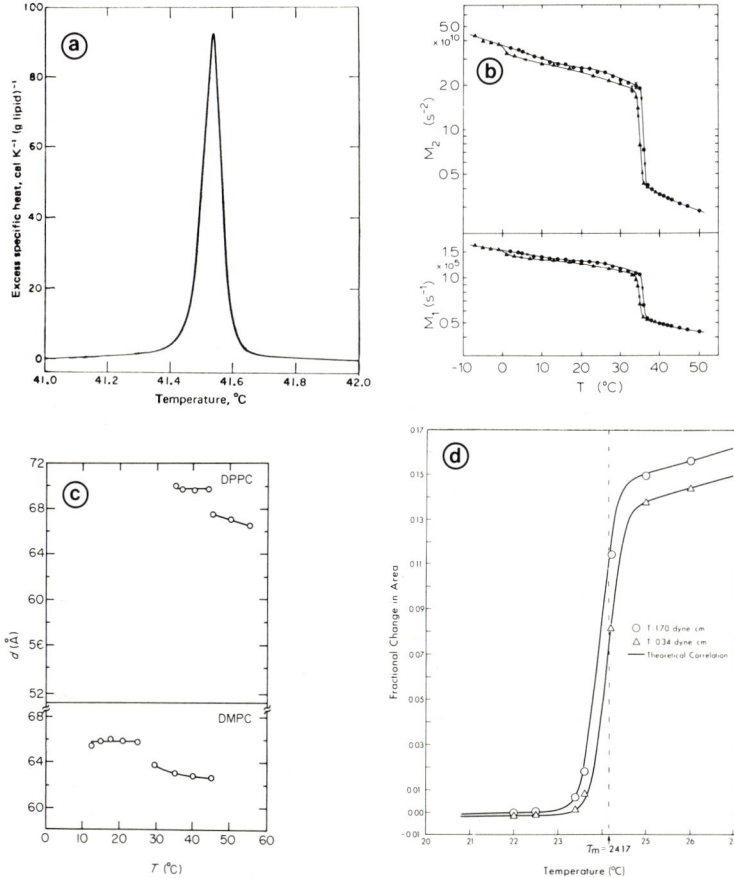

Fig. 4.   Thermal behaviour of one-component lipid bilayers of DPPC and DMPC.  a) Specific heat for DPPC determined by differential scanning calorimetry [After Ref. 29], b) First and second moment of the quadrupole magnetic resonance   spectrum of $d_{62}$-DPPC [After Ref. 30], c) Lamellar repeat distance determined by low-angle x-ray scattering on DMPC and DPPC [After Ref. 31], and d) cross-sectional area change for DMPC determined by micromechanics measurements [After Ref. 32].

where there is a discontinuous change in the multilamellar repeat distance and therefore bilayer thickness (Fig. 4c). The fact that the change in bilayer thickness is
accompanied by a similarly dramatic change in bilayer area is shown in Fig. 4d for
DMPC. Finally, Fig. 4c gives the results for the first two moments of the distribution
of quadrupolar splittings in the nuclear magnetic resonance (NMR) spectrum of perdeuterated DPPC. Again there is a dramatic drop in the moments over a narrow temperature interval [this temperature interval is below that in Figs. 4a and 4c due to the
deuteration]. The moments are an intimate measure of the average segmental order
along the deuterated hydrocarbon chain, i.e.

$$M_1 \sim \sum_i \langle \tfrac{1}{2}(\cos^2 \theta_i - 1) \rangle \qquad\qquad (3)$$

where $\theta_i$, for a deuteron bonded to the ith carbon atom on an acyl chain, is the
angle between the CD–bond direction and the normal to the bilayer.[12] To a good
approximation, $M_1$ is proportional to the hydrophobic thickness of the bilayer.[12,13]

The sharp thermal anomalies shown in Fig. 4 are distinct signals of a phase transition in the bilayer. This transition, commonly referred to as the main transition, or
the gel-to-fluid transition, takes the bilayer from a low-temperature solid gel phase to
a high-temperature fluid phase. Many lipid systems have further bilayer transitions
at lower temperatures, as well as transitions to non-bilayer phases. We shall not be
concerned with these phase transitions here as they are less important for biological
phenomena (except maybe for the polymorphic transformations[34]).

## 2.3 Biological membranes also display lipid phase transitions

Obviously, a biological membrane, being an annealed many-component system,
does not display phase transitions in a strict sense. Rather, one should talk about
phase behaviour and high-dimensional phase diagrams with coexistence regions.
Still, bilayers formed by the lipid extract from a variety of cell membranes exhibit
strong thermal behaviour in a narrow temperature regime in the neighbourhood of
ambient temperatures.

One example is the plasma membranes of the mycoplasma *A choleplasma Laidlawii*[35]
which, as seen in Fig. 5, show strong reminders of a phase transition. Another more
striking example is the thermal anomalies of the hypothalamic phospholipid membranes
of an ectotherm reptile, the garden lizard *Calotes Versicolor*, Fig. 6, which shows that
there is a significant correlation between the 'phase transition temperature' of the
membrane phospholipids and the acclimation temperature. Since a similar correlation
is not found between the acclimation temperature and the 'transition temperature' of
lipids extracted from other parts of the brain, the conspicuous results of Fig. 6 provide a physical basis for the neurophysiological concept that the hypothalamus supports thermoregulatory processes.[37]

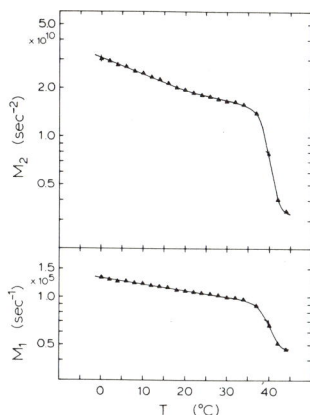

Fig. 5. Temperature dependence of the two first moments of the $^2$H–NMR spectrum of *Acholeplasma Laidlawii* membranes [After Ref. 36].

Fig. 6. Differential scanning calorimetric scans in temperature of the hypothalamic plasma membranes of *Calotes Versicolor* acclimated to different temperatures [After Ref. 37].

There exists no clear-cut evidence that the phase transition phenomena *per se* is important for the physiological functions of membranes, though there are a number of strong suggestions. Again it is important to keep in mind that real membranes are many-component systems with complex phase behaviour and phase diagrams with lateral phase separation regions. Furthermore, biological activity takes place under stable temperature conditions. Membrane phase changes would therefore be of little importance if they could be triggered by changes in temperature only (see, however, Ref. 38). Still, phase changes can also be induced by alterations in chemical composition of the membrane as well as by certain environmental variations, such as changes in pH, ionic strength, and the electric field across the membrane.

Despite the fact that most living organisms function at a fairly constant temperature above the thermally induced lipid membrane phase transition, it is still important to study the response of membranes to changes in temperature. This is the typical physicist's approach to Nature: to make a system reveal secrets of its "natural state" by observing responses to perturbations away from this state.

## 2.4  Fluidity in membrane biology[39]

The garden lizard referred to above and in Fig. 6 evidently prefers its hypothalamic membranes to be in the fluid phase. This complies well with the general observation that biological viability requires the membranes be in a fluid state. Bacterial membranes seem to be close to the transition region whereas mammalian

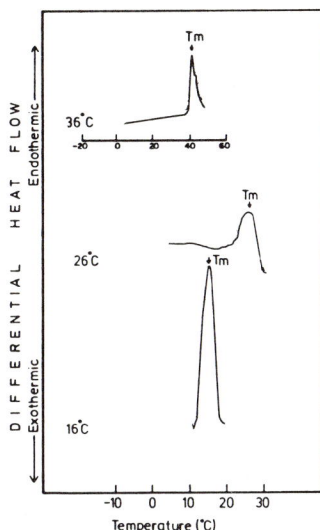

membranes are safely fluid. It is amusing to note that homeothermic animals which hibernate during winter reduce their body temperature, but maintain their membranes in the fluid phase by altering the lipid composition to correspond to a lower transition point.[40] In most cases, membrane-bound enzymatic activity drops dramatically when the membrane enters the gel phase.[41-43] Organisms which live under extreme physical conditions, e.g. thermophilic bacteria[44] in hot springs or deep-sea bacteria[45], retain their membranes in the fluid phase by composing their membranes by either trans-membrane cross-linked lipids or polyunsaturated lipids.

A word of caution should be said regarding the widely used term membrane *fluidity*. Authors sometimes use this term as well as the terms *rigidity* and *micro-viscosity* without adequately defining them. The crux of the matter is that such terms refer to bulk isotropic systems and can only be used by exercising great caution when a highly anisotropic system is in question, such as the thin bimolecular lipid membrane. It is, for example, a common misconception that greater lipid chain orientational order necessarily imparts greater rigidity to the membrane matrix. The influence of cholesterol on membrane chain order and lateral mobility as discussed in Sec. 3.6 demonstrates that this is not generally true. Basically, membrane fluidity is determined by two components, lipid chain orientational order and microviscosity. For the purpose of relating biological activity to a membrane physical property, the lateral and rotational molecular mobility or diffusion, i.e. microviscosity, seems to be the appropriate physical property to consider.[46]

## 2.5 Basic experimental observations for lipid bilayer phase transitions: Lipid chain melting

The main gel-to-fluid lipid bilayer phase transition, cf. Fig. 4, is associated with the following characteristics: i) It is a sharp endothermal first-order transition (at a temperature $T_c$), ii) it is accompanied by a large area expansion, typically $\Delta A(T_c) \lesssim 20\%$, iii) it involves hardly any volume change,[47] $\Delta V(T_c) < 4\%$, and iv) it leads to a large transition entropy, $\Delta S(T_c) \sim 15k_B$/molecule. The approximate conservation of volume implies that there is a constant reciprocal relationship between lipid bilayer thickness and cross-sectional area per molecule.[18] Thus, as Fig. 4 also shows, the bilayer gets thinner during the transition.

A simple estimate, using the Bolzmann formula, shows that a very large number of degrees of freedom, $\exp(\Delta S/k_B) \sim 10^6$, is activated during the phase transition. Thus the lipid bilayer phase transition is orders of magnitude more entropic than ordinary three- or two-dimensional melting transitions or smectic melting. The source of this entropy is the internal degrees of freedom of the lipid hydrocarbon chains, i.e. the rotational isomerism. Hence, the lipid bilayer transition is reminiscent of polymer melting. This has led to the term *chain melting* to characterize the lipid bilayer phase transition.

## 2.6   Models of lipid membrane phase transitions

It has been a substantial challenge to physicists to model the gel-to-fluid lipid
bilayer phase transition and to identify the relevant symmetries and interactions.  The
early theories have been critically reviewed by Nagle[19] and we shall here concentrate
on some more recent approaches which seek to account for the finer details of the
transition.

A proper theoretical model must take into account the following basic physics of
a condensed wet lipid aggregate: i) the rotational isomerism of the lipid chains,
ii) the anisotropic van der Waals forces between the hydrophobic part of the lipids,
iii) the polar forces between the head groups, iv) the excluded volume interactions,
and v) the interaction with water.

Different degrees of realism can be imparted to a theoretical model.  However,
it is quite clear that a highly realistic model will imply high complexity and thus make
difficult a detailed calculation of the properties of the model.  In the present stage of
the theory of biomembrane phase transitions, where the relevant interactions mentioned
above have been identified using highly simplified models which in some cases allow for
exact calculations,[19] there is a call for models with a high degree of realism which are
able to measure up to the advanced and detailed level of the experimental programmes.
The complexity of such models has made it necessary to use modern computer simulation
techniques[22] to derive the properties of the models.  A number of molecular dynamics
studies of bilayer models with accurate potentials formulated in terms of spatial coordinates have been carried out.[21,48]  Due to the limited number of molecules which
can be treated in this realization, no quantitative information has been obtained in the
transition region.

Consequently, our main information on lipid chain melting in theoretical models
derives at present from somewhat simpler lattice models.  A class of such models, the
so-called multi-state lattice models, has been studied extensively by Pink and collaborators.[20,22,49-53]  Using mean-field theory as well as computer simulation techniques,
this class of models has been found to accurately describe a great variety of bulk
thermodynamic as well as spectroscopic experimental results for phase transitions in
lipid bilayers composed of different lipid species.  In their original formulation the
multi-state models were only incorporating the internal degrees of freedom of the lipid
hydrocarbon chains and neglected, via the lattice formulation, crystallization phenomena in the two-dimensional translational coordinates.  The neglect of translational
degrees of freedom was justified by Doniach's estimate[54] that their contribution to the
enthalpy of transition is approximately 1 kcal/mol and independent of chain length,
whereas the total enthalpy of transition is chain-length dependent and as much as
8.7 kcal/mol for DPPC multi-bilayer systems.  A fundamental question now arises:
does the bilayer condense in terms of the translational and internal degrees of freedom
at the same time and how does the two-dimensional crystallization process affect the

bilayer properties. Recent wide-angle x-ray synchrotron studies of certain *in situ* lipid monolayers indicate that the transitions in terms of the two sets of degrees of freedom may indeed be decoupled.[55,56]

The question of the effects of acyl chain ordering as well as two-dimensional crystallization on the main phase transition of wet lipid bilayers has recently been approached by means of a generalized multi-state model which incorporates intra-chain flexibility as well as crystallization variables.[52] The model takes into account the internal conformational states of the acyl chains and their mutual interactions, as well as the large number of in-plane orientations of two-dimensional lipid bilayer crys-tallites. To illustrate the use of microscopic interaction models, statistical mechanics, and computer simulation in membranology, we shall describe in some detail the strat-egy and results of this approach.

The model Hamiltonian of the bilayer aggregate consists of two parts[52]

$$H = H_L + H_P , \qquad (4)$$

where $H_L$ is the Hamiltonian for Pink's ten-state model.[14] Within the scope of this model, the bilayer is considered as two monolayers which are independent of each other. Each monolayer is represented by a two-dimensional triangular lattice with $N = L \times L$ lattice sites. Every site of the lattice is occupied by a single saturated hydrocarbon chain which is in one of ten distinct conformational states. Each state is characterized by an internal energy $E_n$, a cross-sectional area $A_n$, and a degeneracy $D_n$, where $1 \leq n \leq 10$. All ten states are derivable from the all-*trans* state in terms of *trans*-gauche isomerism. The two key conformational states are the non-degenerate gel-like ground state ($n = 1$) representing the all-*trans* conformation and a highly degenerate excited state ($n = 10$) characterizing the "melted" or fluid phase. The model is completed by including eight intermediate gel-like states which contain kink and jog excitations. These intermediate states are selected subject to the require-ment of low conformational energy and optimal packing. The values of $E_n$ are deter-mined from the energy required for a gauche rotation ($0.45 \times 10^{-13}$ erg) relative to the all-*trans* conformation. The values of $D_n$ are obtained from combinatorial considera-tions and $A_n$ is calculated using the geometrical constraint that the volume of an acyl chain is invariant.[18] The chains interact via anisotropic forces which represent both van der Waals and steric interactions. The lattice approximation automatically accounts for the excluded volume effects and to a rough approximation for the part of the inter-action with water which brings the bilayer into existence. An internal lateral pressure is added to model the interfacial forces required for bilayer stability.

The second term, $H_P$, of the Hamiltonian is a modification of the high-q-state Potts model, where q is the number of Potts states. The standard-q-state Potts model is a lattice model which has been successfully used to describe grain growth in poly-crystalline aggregates.[57]

The standard Potts model accounts for the grain-boundary energy of a metastable distribution of crystalline domains, each of which is characterized by a Potts state. In the modified Potts model, only the first nine conformational states carry a Potts variable which then describes the orientation of the crystalline domain with which the chains are associated. When the conformational state of a chain changes from a gel-like conformer to the excited (10th) state, it loses its Potts variable, which gives rise to zero grain-boundary energy. This is reasonable since the excited state is representative of the fluid phase of the bilayer, which cannot be in a granular configuration. The grain-boundary energy is modelled by allowing neighbouring acyl chains to interact with an energy $J_P > 0$ if they are in different Potts states. Otherwise the interaction is zero. Since $H_P$ describes crystallization on an underlying lattice by means of Potts variables which only indirectly take account of translational degrees of freedom, it is of course unable to describe more subtle two-dimensional effects, such as "two-dimensional melting" and the possible occurence of a "hexatic" phase.[58]

The phase behaviour described by the conformational Hamiltonian, $H_L$, is well understood.[49-51] Some results from Monte Carlo computer simulation studies of the ten-state model are shown in Figs. 7 and 8. Figure 7 demonstrates that the gel-to-fluid transition within this model is signalled by dramatic changes in cross-sectional area and internal energy of the chains. The numerical observation of a "continuous

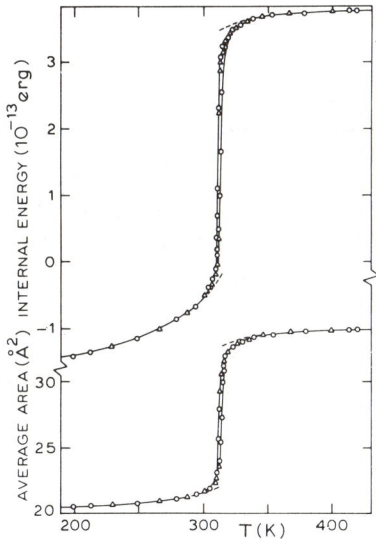

Fig. 7. Temperature dependence of internal energy and cross-sectional area per chain for a ten-state model of lipid bilayers. Model parameters are pertinent to DPPC [After Ref. 49].

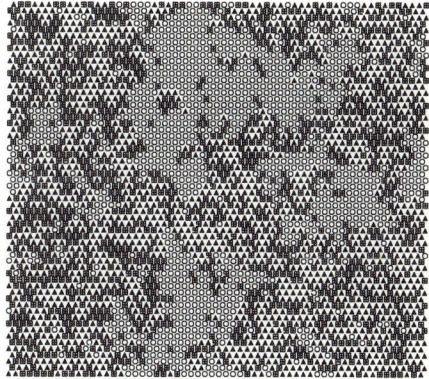

Fig. 8. Snapshot of a micro-configuration of a multi-state lipid bilayer model at a temperature very close to the equilibrium phase transition temperature. Triangles indicate gel chains, circles fluid chains, and squares intermediate chain states [After Ref. 51].

transition" with a slight hysteresis offers an explanation of the corresponding experimental finding (cf. Fig. 4) : rather than being ascribed to the presence of impurities it is a result of a kinetically caused metastability of intermediate lipid-chain conformations which facilitate the formation of long-lived clusters in the transition region, Fig. 8. That the equilibrium transition is indeed of first order is documented by the free energy curves in Fig. 9. The apparent continuous nature of the transition has been

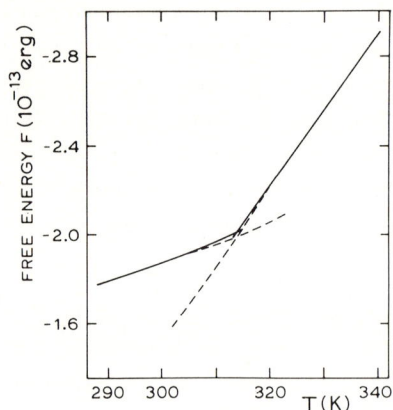

Fig. 9. Free energy vs temperature for the ten-state model of lipid bilayers. Model parameters are for DPPC [After Ref. 49].

described in terms of pseudo-critical phenomena.[51,59,60] The kinetic behaviour of the phase transition of $H_L$ is less well understood[22] although there are strong indications that the transition is characterized by several relaxation times, in accordance with recent dynamic fluorescence experiments on DPPC vesicles.[61]

Inclusion of the modified Potts Hamiltonian in Eq. (4) changes the equilibrium phase behaviour and leads to the qualitative phase diagram shown in Fig. 10. Three distinct phases occur: 1) a crystalline gel phase, 2) a disordered gel phase, and 3) a fluid phase. It is anticipated, in accordance with recent combined low- and wide-angle x-ray synchrotron studies of multilamellar lipid dispersions[62] and kinetic calorimetric measurements,[63] that values of $J_P$ realistic for bilayers will be above the triple point value in Fig. 10. In contrast, some monolayers may indeed possess the structurally disordered solid gel phase.[55,56] A striking prediction following the coupling of the two transitions is that the non-equilibrium behaviour of the bilayer transition is characterized by interfacial melting,[52] a process by which the grains of the low-temperature polycrystalline bilayer melt inwards from the boundaries. This phenomenon which still awaits an experimental verification is pictured in Fig. 11. It is closely related to the phenomena of wetting and interfacial adsorption.[64]

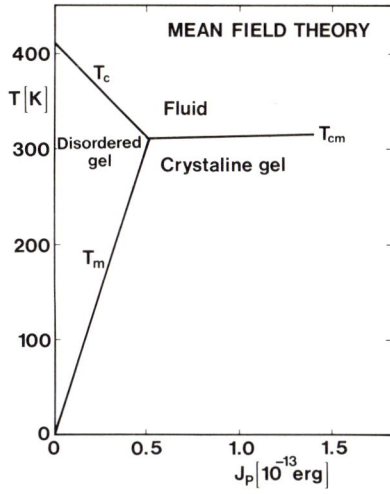

Fig. 10. Mean-field phase diagram
for the lipid bilayer model in Eq. (4)
[After Ref. 52].

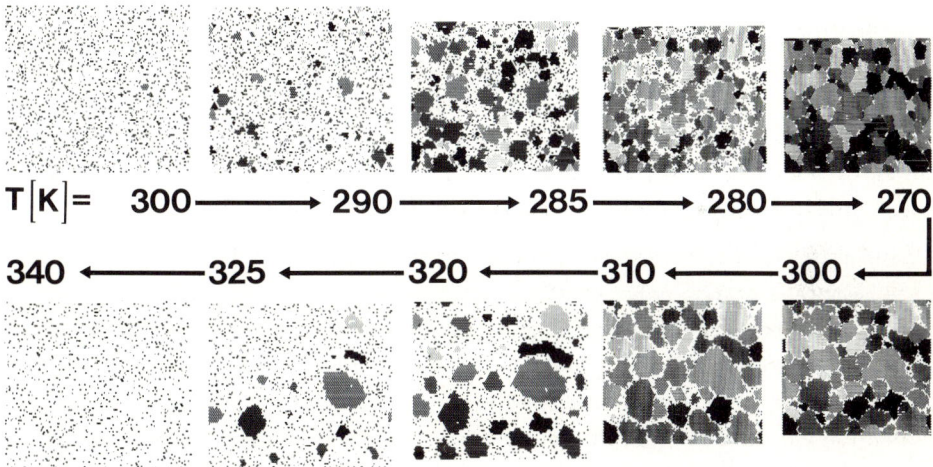

Fig. 11. Snapshots of microconfigurations illustrating the non-equilibrium
processes of grain boundary formation and subsequently interfacial melting
for a lipid bilayer model studied by computer simulation [After Ref. 52].

## 3. EFFECTS OF 'IMPURITIES':
## PROTEINS, POLYPEPTIDES AND CHOLESTEROL

### 3.1 Membrane-bound proteins: Structure and function

Compared to water-soluble proteins, the elucidation of the three-dimensional structure of integral membrane bound proteins has progressed much more slowly.[16] The reason for this is a physical one: In order to be accommodated in a lipid bilayer membrane, an integral protein has to be amphiphilic with a hydrophobic region between the hydrophilic ends which are anchored in the membrane polar head group region, cf. Fig. 1. Hence, it is difficult to isolate and purify such proteins in order to form three-dimensional crytals which are prerequisites for a proper structural investigation by means of scattering techniques. Conversely, the primary amino acid sequence of an increasing number of membrane proteins becomes available. On the basis of such sequences it is in some cases possible, on the basis of energetic arguments related to measures of hydrophobicity, to infer gross details of secondary structure (e.g. content of $\alpha$-helices or $\beta$-sheets) and tertiary structure.[15,65] Still, the fundamental problem of protein folding is largely an unsolved one. It is imporant to realize that the part of the protein buried in the hydrophobic interior is "inside-out" in the sense that hydrophobic residues will be abundant on the exterior surface of the protein, whereas there may very well be hydrophilic and even charged residues buried inside the protein (the reverse is true for water-soluble proteins). In fact, such buried hydrophilic and charged residues may support protein functions related to transport of ions and hydrophilic molecules. If integral membrane proteins are removed from the membrane or a membrane-like environment (e.g. provided by an appropriate detergent solution), they usually denaturate and change their structure.

Until recently the three-dimensional structure had been worked out for a single integral membrane protein only, bacteriorhodopsin, which acts as a light-sensitive proton pump in the purple membranes of the microorganism *halobacterium halobium*.[15,16,66] It is a pure fluke of Nature that this protein spontaneously forms two-dimensional crystalline arrays in its own natural membrane, thus allowing, via ingenious image-processing of electron micrographs, a reasonably resolved (5-10Å) three-dimensional picture to be drawn.[16] Figure 12 shows the three levels of structural characterizations of bacteriorhodopsin. The hydrophobic core is seen to consist of seven predominantly $\alpha$-helical polypeptide sequences[67] which traverse the membrane. At the membrane interface these segments are bound together two and two by segments of hydrophilic residues. There is reason to believe that the overall structure of bacteriorhodopsin is so strongly dictated by the general nature of the lipid membrane matrix that this structure is the prototype of integral membrane proteins.[15]

Important progress has recently been made in developing methods to crystallize integral membrane proteins by performing complex-formation with specific ligands.[68]

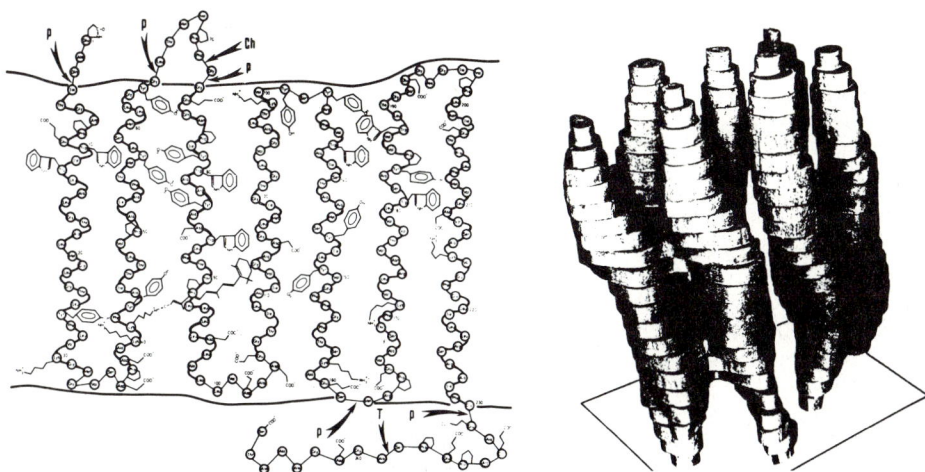

Fig. 12. A typical integral membrane protein: bacteriorhodopsin. To the left are shown the primary amino acid sequence and secondary structure in an arrangement which matches the hydrophobic and hydrophilic regions of the membrane. To the left is given a three-dimensional model of the hydrophobic, predominantly α-helical, membrane-spanning segments [After Ref. 15].

This led to the first x-ray high-resolution ($3\overset{o}{A}$) structure of an integral membrane protein, the photosynthetic reaction centre of *rhodopseudomonas viridis*.[17,69] It was indeed found that the two protein subunits of the reaction centre each consist of a number (five) membrane-spanning α-helices, in accordance with the statement above that the overall structure of membrane proteins is subject to some universality.

A fundamental question is whether there is a relationship between function of membrane-bound proteins and enzymes – and the structure, dynamics and thermo-dynamics of the lipid bilayer membrane. Specifically, how do the proteins affect the lipid properties and, conversely, how do the lipids modulate the protein function? Related to this are questions regarding lipid specifity[41] and the functional roles of lipids.[34] A lot of research is still needed to give definite answers to these questions. However, it is anticipated that membrane structure and thermodynamics form the phys-ical basis of trigger processes in membranes.[70]

## 3.2 Lipid-protein interactions[71]

Very little is known at present regarding internal molecular conformational states of integral membrane proteins and their relation to modes of protein activity. The bulk of our quantitative information on the physical effects of the interaction between lipids and proteins is concerned with the possible perturbing effects of proteins on their lipid environment. In addition, substantial information is available about lipid-mediated protein-protein interactions and their consequences in terms of protein aggregation. We shall not attempt to review the vast literature which now exists in this field, but rather refer to the articles in two recent books.[71] Neither shall we discuss the almost classical controversy in interpreting effects of lipid-protein interactions as they manifest themselves in different experimental techniques, e.g. nuclear and electron spin resonance.[11,72] Time-scale arguments seem to have resolved at least part of this controversy by now.

Instead we shall focus on the striking experimental observation from NMR spectroscopic studies of lipid-protein recombinants that, whereas the lipid chain orientational order is progressively decreased when proteins are incorporated in the gel phase, there is hardly any change in orientational order in the fluid phase. An example of this almost universal phenomenon is given in Fig. 13 in the case of rhodopsin - DMPC

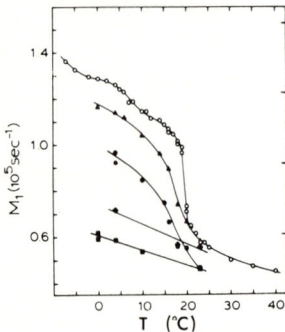

Fig. 13.  First moment vs temperature of the quadrupolar splitting for rhodopsin - DMPC recombinants for different protein concentrations.  In the gel phase, $M_1$ decreases with increasing concentration.  [After Ref. 73].

recombinants.[73] It is noticed that the order is only slightly affected in the fluid phase and that the order in the gel phase decreases with increasing protein concentration and eventually at high protein contents reaches the level of the fluid phase.[74] The effect appears even more striking when compared with the results of a similar experiment on a lipid-cholesterol membrane. For such a system one also finds decreasing order in the gel phase, but a dramatic increase of lipid order in the fluid phase (see also Sec. 3.6).[75] Why is this so, and does it tell us something important about the fundamental and relevant features of the interaction between lipids and proteins?

## 3.3  Models of lipid-protein interactions

The apparent universal properties of integral membrane protein structure hold a promise for the possibility of constructing a general theory of lipid-protein interactions seeking to encompass features which a large class of membranes will have in common. This is particularly fortunate considering the current limited knowledge of details in protein structure.

Theories of lipid-protein interactions have recently been reviewed extensively by Abney and Owicki.[23] We shall here restrict ourselves to outlining the general strategies adopted by the theorists, and describe in some detail one of the approaches which pays special attention to the role of hydrophobic matching.

From the standpoint of statistical mechanics, the influence of proteins on the phase behaviour and thermodynamic properties of the lipid matrix has been studied along two different routes. One is by means of phenomenological Landau theories which have proven useful in describing phase transitions in a great variety of other physical systems. In these theoretical developments, the dependence of various thermodynamic quantities such as transition temperature and heat of melting on protein concentration has been described qualitatively. Special attention is given to how the perturbation of the lipid order decays away from the surface of the individual proteins as expressed in terms of a coherence length for spatial fluctuations. In all cases, the protein is treated formally as a rigid boundary condition for the lipid order parameter. It has been shown that quantities such as heat capacity and lateral compressibility, which are closely associated with fluctuations are enhanced near the phase transition and that the membrane may be driven towards a "critical" point when the protein concentration is increased.[76] A common feature of the Landau theories is that they are formulated in terms of mainly unknown expansion parameters which are difficult to relate to measurable physical properties. More important, these approaches have assumed homogeneous dispersions of proteins and have therefore excluded, à priori, the possibility of lateral phase separation which is known to occur in most lipid-protein mixtures, especially in the gel phase. The other route taken by theories of lipid-protein interactions has involved the use of detailed microscopic statistical mechanical models previously developed to describe the main gel-fluid transition in pure lipid bilayers by taking explicitly into account the nature of the molecular interaction forces as well as the statistics of the hydrocarbon chain conformations. In Marčelja's model,[77] the protein is introduced as a cylindrical boundary condition on the lipid orientational order, and non-specific lipid-protein interactions are assumed. The model does not allow for phase separation and the derived results are very similar to those obtained from the Landau-type theories. In the approach by Pink and co-workers,[78] a much more detailed microscopic model is drawn upon including specific lipid-protein interactions which depend on the configurational state of the individual hydrocarbon chains. The Pink model has proven useful in describing a variety of experimental observations for different lipid-protein mixtures. The basic drawback of such microscopic model calculations is the introduction of a large number of model parameters which can only be determined by elaborate fitting to experimental data.

With the purpose of describing physical theories of lipid-protein ineractions in membranes in more detail, we shall discuss a particular model, the mattress model,[33] which has been designed specifically to elucidate the concept of hydrophobic matching of membrane-spanning amphiphilic molecules such as polypeptides and integral proteins. In the spirit of a physicist's approach, this model examines the potential importance of hydrophobic match as the 'natural state' of membranes by examining effects of a hydrophobic mismatch. The mattress model visualizes the membrane as an elastic sheet ('the mattress') with imbedded stiff particles ('the springs') perturbing the membrane hydrophobic thickness. The model has a number of features in common with elastic theories advanced to describe staging in intercalation compounds.[79] The energy stored in the undulations of the membrane surface caused by the mismatch is related, within the model, to the elastic properties of the lipids and proteins. In contrast to the micro-scopic theories mentioned above, the mattress model makes no attempt to describe the properties of the pure lipid system itself, but accepts as input data the known thermo-dynamic properties of the pure lipid bilayer, including the properties of the phase transition, and seeks only to model the perturbations produced by the proteins. In addition to the elastic distortion forces, the model incorporates indirect lipid-protein interactions induced by the mismatch, as well as direct lipid-protein van der Waals-like interactions between the hydrophobic parts of the lipid bilayer and the proteins. The model is solved within the framework of a two-component real solution theory which includes the possibility of phase separation.

With reference to the schematic drawing of the mechanical analogue of the mattress model pictured in Fig. 14, the free energy of the model can be expressed

$$G = G_{ideal} + H_E$$

$$= G_{ideal} + X_L A_L (d_L - d_L^0)^2 + X_L X_P B_{LP} |d_L - d_P| + X_L X_P C_{LP} \min \{d_L, d_P\}, \qquad (5)$$

where $G_{ideal}$ is the free energy of an ideal mixture and where it has been assumed that the protein is rigid. $A_L$ is the elastic constant of the lipid bilayer (or the inverse

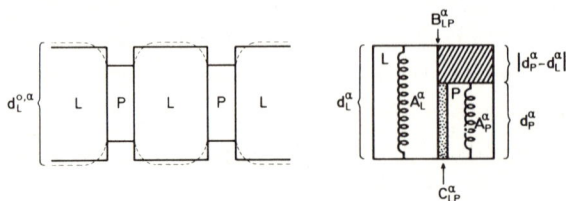

Fig. 14. Mechanical analogue of the mattress model of lipid-protein interactions in membranes. Only the hydrophobic regions of the lipid bilayer (L) and the amphiphilic protein or polypeptide (P) are indicated. The cross-hatched area corres-ponds to hydrophilic regions (water or head groups) and the dotted region represents the adhesive hydrophobic dispersion force [After Ref. 33].

lateral compressibility) and $B_{LP}$ and $C_{LP}$ are interaction constants related to the hydrophobic effect and the hydrophobic van der Waals forces, respectively. $X_L$ and $X_P$ are concentration variables and $d_L$ and $d_P$ are the fundamental geometrical variables, the hydrophobic thicknesses, of the theory. $d_L^0$ refers to the pure bilayer thickness. Using an expression of the type Eq. (5) for each lipid bilayer phase, the phase diagram spanned by $X_P$ and temperature can be determined. Figure 15 shows the qualitative outcome of such a determination for three typical situations: a) the protein is more soluble in the gel phase, b) the protein is more soluble in the fluid phase, and c) the protein is more soluble in the gel phase at low concentrations. An important part of the

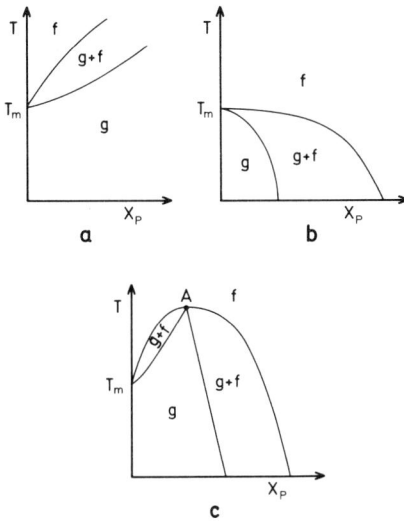

Fig. 15. Phase diagrams of the mattress model of lipid-protein inter-actions in membranes. f and g refer to fluid and gel phases and g +f to the coexistence region [ After Ref. 33].

solubility, i.e. the chemical potential of the protein in the appropriate phase, is the degree of hydrophobic matching. Case c) is particularly interesting: The protein best matches the gel lipid bilayer hydrophobic thickness and the initial phase boundaries will be pushed upwards. However, incorporation of protein in the gel matrix costs elastic energy which, since $A_L^g > A_L^f$, is larger than the corresponding cost in elastic energy in a fluid matrix. Thus there will be a competing effect which will eventually take over and reverse the trend of the phase boundaries revealing an upper melting point (an azeotropic point).

The mattress model suggests that, in the case of hydrophobic matching, the lipid membrane will be unperturbed. Returning to the experimental observation of Fig. 13 and noting that the average NMR orientational order parameter is proportional to bi-layer hydrophobic thickness,[33] it can be argued that the lack of perturbation due to lipid-protein interactions is caused by an approximate matching of lipid and protein hydrophobic thicknesses. The reason for this is obviously the biophysicists' pre-occupation with certain proteins and lipids related to natural membranes. Support for this point of view is provided by a recent experiment on reaction centre proteins of *rhodopseudomonas sphaeroides*[80] reconstituted in phosphatidylcholine vesicle mem-branes of different thicknesses, where it was found that the shift of the midpoint transition temperature depends on the hydrophobic matching, cf. Fig. 16.

Similarly, the mattress model provides an appropriate framework for explaining calorimetric experimental data for a large selection of lipid-protein membranes with respect to phase diagram and enthalpy of melting as a function of protein concentra-tion.[33] Also the event of protein aggregation as well as which lipid environments may facilitate aggregation can be studied within the mattress model.[33] The driving force for aggregation within the mattress model is provided by the lipid-mediated indirect protein-protein attractive interaction which lowers the elastic energy, but competes with the entropy of mixing. More realistic models of protein aggregation have to include information on direct protein-protein interactions and the chemical potential of bulk protein phases.

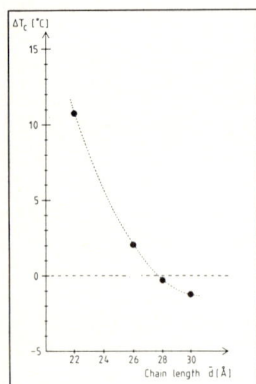

Fig. 16. Shift of midpoint transition temperature as a function of mean membrane thickness,

$$\bar{d} = (d_L^g + d_L^f)/2$$ for photosynthetic RC protein $(X_P = 10^{-4})$ reconstituted in phosphatidylcholine membranes of different thicknesses [After Ref. 80].

## 3.4 Polypeptides as model proteins

The interpretation of experiments on lipid-protein membranes in terms of theoretical models is severely hampered by a lack of quantitative characterization of the state of protein aggregation in the lipid membrane. There are some indications from freeze-fracture studies that most proteins, except at very low concentrations, form aggregates or solid patches in the lipid gel phase. Thus, the phase diagram may be much more complicated than anticipated by Fig. 15, e.g. including eutectic behaviour.[33] On top of this, the overall geometry of many integral membrane proteins is little known.

This situation has stimulated work on simpler systems of synthetic lipid membranes with model proteins, e.g. ion channels and polypeptides. An intensively studied cation channel is gramicidin A[81] whose structure is known. A particularly interesting and very promising class of model systems makes use of synthetic amphiphilic polypeptides synthesized specifically to span lipid membanes.[82] Use of such polypeptides in an α-helical configuration to imitate important physical aspects of proteins is justified by the observation that membrane-spanning α-helical polypeptides are abundant elements of integral membrane proteins.

Spectroscopic and thermodynamic studies [82-84] have been carried out on polypeptides of the type

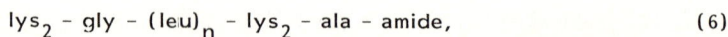

$$\text{lys}_2 - \text{gly} - (\text{leu})_n - \text{lys}_2 - \text{ala} - \text{amide}, \tag{6}$$

where the n leucines, n = 16, 20, 24, form an α-helix in the hydrophobic core of a lipid bilayer. That the mattress model is indeed useful in rationalizing the effect of lipid-polypeptide interactions on the basis of hydrophobic matching is indicated by the experimental results given in Fig. 17 which are obtained for perdeuterated DPPC membranes incorporated with polypeptides of n = 16 or n = 24 leucine residues. The short

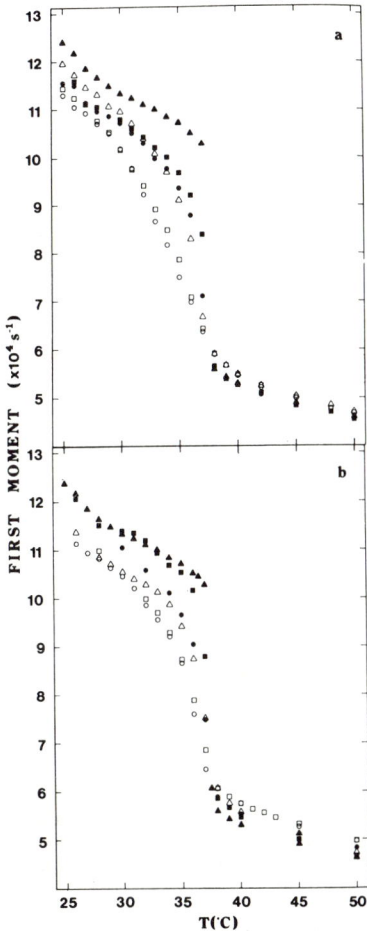

Fig. 17. Temperature dependence of the first moment of the deuterium NMR quadrupolar spectrum of DPPC lipid bilayers with polypeptides of the type in Eq. (6): a) n = 16, and b) n = 24. Increasing polypeptide concentration from $X_P$ = 0 to 0.02 leads to decrease of chain orientational order in the gel phase [After Ref. 83].

peptide has $d_P$ = 23Å and matches rather closely the fluid lipid DPPC bilayer, $d_L^{o,f} \simeq$ 26Å, whereas the long peptide, $d_P$ = 35Å, is in between $d_L^{o,f}$ and $d_L^{o,g} \simeq$ 42Å (corrected for the tilt).[33] As expected on the basis of the mattress model predictions, both peptides perturb the gel phase dramatically and the perturbation increases with $X_P$. Conversely, the short peptide only affects the fluid bilayer thickness very slightly, whereas the long peptide increases the bilayer thickness progressively with increasing $X_P$. Referring back to the experimental observations on rhodopsin in lipid bilayers, Fig. 13, these experiments on simple and well-defined model membranes lend evidence to the belief that hydrophobic matching is relevant for lipid-protein interactions in membranes.

## 3.5   Membrane thickness, hydrophobic matching, and physiological function

It is now suggested that hydrophobic matching between proteins and lipids is required for optional functioning of membranes.  Strong support for this statement comes from a number of investigations, of which we shall describe a few.  Figure 18 gives the measured lifetime of the dimer channel, gramicidin A, incorporated in mono-acylglycerol bilayers of different thicknesses.[85]  It is seen that the lifetime is at its optimum around a bilayer hydrophobic thickness which is close to the hydrophobic

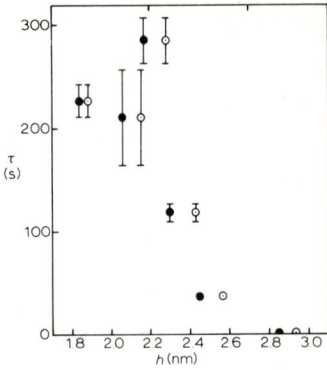

Fig. 18.   Gramicidin A channel lifetime as a function of thickness of the hydrocarbon region of lipid bilayers [After Ref. 85].

length of the channel, $d_P \simeq 22\overset{o}{A}$.  Non-matching bilayers obviously introduce a tension on the channel and lower the lifetime.  Similarly, nerve impulses can be blocked by anaesthesia, such as n-alkanes, as illustrated in Fig. 19 which shows that the action potential is diminished after introducing n-alkanes into squid giant axons.[86]  It is known

Fig. 19.   Nerve impulse blockage by n-alkanes, $C_n$, in squid giant axons [After Ref. 86].

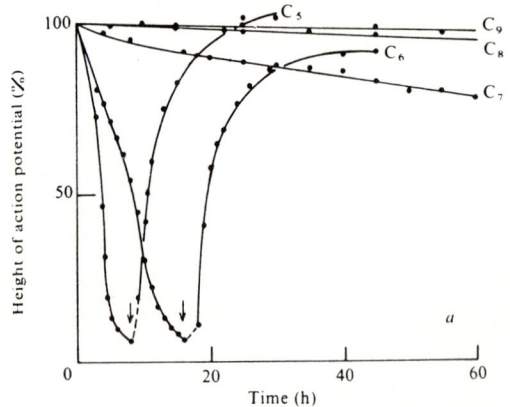

that n-alkanes become dissolved in the hydrophobic interior of lipid membranes and effectively increase the hydrophobic thickness.[87]  Obviously, the thicker membranes destabilize the ion-channels in the axon membranes, causing a drop in the action potential. Alkanols are known to lead to a similar blocking. In the light of the thick-ening effect of cholesterol in fluid membranes, it is also amusing to note that incorpora-tion of cholesterol in the neural membranes of *aplysia californica* decreases its electrical activity.[88]

Turning now to proteins and enzymes, Fig. 20 shows that the integral membrane protein cytochrome-c-oxidase exhibits an optimum in its activity as a function of the thickness of the lipid bilayer membrane in which it is reconstituted.[89] Similar optimums are found for enzymatic activities of $(Na^+ - K^+)$ - ATPase[90] and $(Ca^{2+} - Mg^{2+})$ - ATPase [91] reconstituted in monounsaturated (n:1) phosphatidylcholine bilayers of different

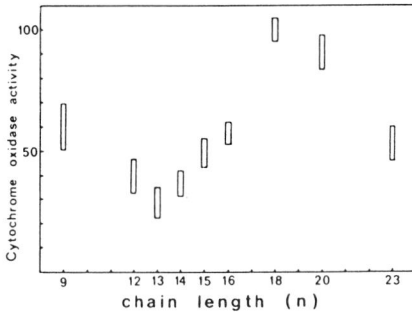

Fig. 20. Effect of phosphatidylcholine hydrocarbon chain length on the activity of bovine heart mitochondrial cytochrome-c-oxidase [After Ref. 89].

thickness, cf. Fig. 21. It is anticipated that intercalation of hydrocarbons into lipid bilayers imparts greater 'fluidity' to the membrane. The fact that it is not the degree of 'fluidity' as such, but rather hydrophobic matching which stimulates the enzymatic activity in fluid bilayers is very convincingly demonstrated by the experiments repres-ented in the right hand side of Fig. 21. In these experiments the bilayer thickness is

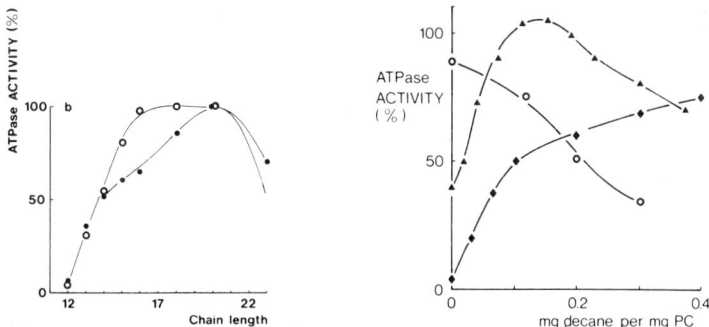

Fig. 21. Left: Activity of $(Na^+ - K^+)$ ATPase (o) and $(Ca^{++} - Mg^{++})$ ATPase (●) as a function of lipid chain length at 30°C. Right: The effect of n-decane on the activity of $(Na^+ - K^+)$-ATPase in bilayers with different lipid chain lengths: (12:1) diamonds, (14:1) triangles, and (20:1) circles [After Ref. 90].

manipulated by incorporating n-decanes in the bilayer. It is found that the lost enzym-atic activity in too thin bilayers can be regained by increasing the thickness via hydrocarbons, whereas the low activity in too thick bilayers is lowered even further upon the addition of decane. The full scenario of these effects is illustrated in Fig. 22.

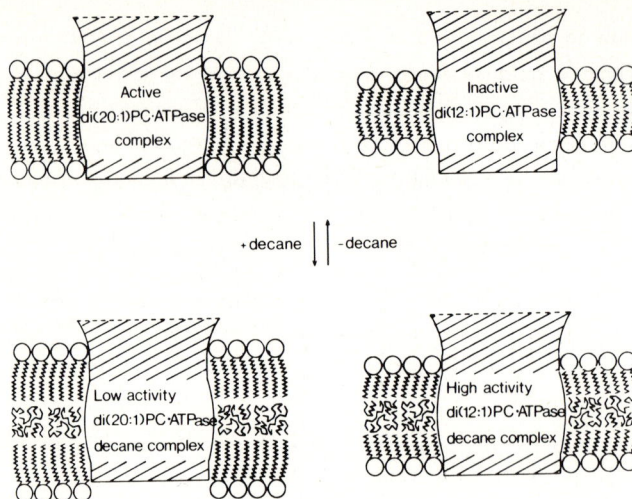

Fig. 22. Mechanisms of deactivation and reactivation of ATPase in lipid bilayers of different hydrophobic thicknesses. The thickness can be manipulated by intercalating n-decane [After Ref. 91].

To summarize, we find that there is ample theoretical and experimental evidence in favour of hydrophobic matching being an important parameter for characterizing the effects of lipid-protein interactions in membranes. The mattress model described in Sec. 3.3 provides a convenient framework within which such effects can be rationalized. Furthermore, and equally important, it provides a simple basis for visualizing and suggesting a quantitative programme geared towards a fuller understanding of membrane structure and function.

## 3.6 Cholesterol function in lipid membranes[92,93]

Sterols are abundant in many biological membranes. In mammalian cells the major sterol component is cholesterol, whereas ergosterol occurs frequently in plant cell membranes. Cholesterol is a truly unique molecule which Nature took great care as well as a long time to evolve (cf. Sec. 3.8). The molecule has a rigid sterol skeleton and a little, flexible hydrocarbon tail, cf. Fig. 2. It has a high degree of hydrophobic smoothness. Cholesterol is the terminus of a long biochemical pathway, starting with non-cyclic squalene produced in an anaerobic environment. In the presence of molecular oxygen, squalene becomes aerobically cyclicized to lanosterol, which is then the immediate cholesterol precursor.[94]

Cholesterol has some dramatic effects on membrane physical properties: a) The lipid chain order is decreased in the gel phase and increased in the fluid phase. For

high cholesterol contents, the gel-to-fluid phase transition is eliminated.[75] b) No surface shear restoring forces are operative over a wide temperature range, indicating
fluid-like macroscopic properties.[95] c) Translational and rotational diffusion constants
are not strongly affected in the high-temperature region, indicating a fluid-like microviscosity.[96] d) The lateral compressibility decreases, indicating greater mechanical
rigidity.[97] e) The transmembrane permeability is lowered, indicating a sealing effect
of cholesterol. The implication of a) is that cholesterol thickens fluid lipid membranes.

Thus cholesterol has the unique capacity of mechanically strengthening ('rigidify-
ing') fluid lipid membranes without changing the microviscosity component of the bilayer fluidity. Probably very few, if any, other amphiphilic molecules have a similar
capacity.

We have been concerned above with the pure physical effects of cholesterol in
membranes. It should be noted that cholesterol is also known to play an important
role in metabolic functions.[94] Many membrane receptors are subject to control of
cholesterol (possibly the acetylcholine receptor) and cholesterol also plays a central
role in the regulation of membrane biosynthesis, cell growth and cytosis.[93] The interaction between cholesterol and protein function has no general character: cholesterol
may stimulate, inhibit, or have no effect at all.[93] It has a well-documented indirect
effect on proteins in that it promotes protein aggregation in fluid membranes,[97] possibly via the hydrophobic thickening and mismatch effect discussed in Sec. 3.5.

Only very few quantitative theoretical studies have been carried out on lipid-
cholesterol interactions, and the situation regarding the theoretical understanding of
the complex lipid-cholesterol phase diagram is highly unsatisfactory. We refer to the
theories described in Refs. 98-101, 24 and 49. Most of these theories treat lipid-
cholesterol interactions in a microscopic framework, including specific interactions between the rigid cholesterol molecule and certain excited lipid chain conformational
states.

## 3.7   Other agents interacting with lipid membranes

Numerous chemical species are known to have significant effects on biomembrane
functions. In fact, a large fraction of pharmaceutical substances or drugs have the
membrane and its proteins and receptors as their targets. We shall restrict ourselves
here to giving a brief and by no means complete list of some agents which influence the
physical properties of membranes.

Fatty acids may alter the order in the polar head group region, as well as in the
hydrophobic interior, depending on the fatty acid chain length and the degree and type
of unsaturation.[102] Various antibiotics, such as filipin,[103] amphotericin B,[104] and
alamethicin[105] produce major changes in the membrane permeability in the presence of
cholesterol and promote leakage and ultimately lysis of cells. Such effects can be explained on the basis of packing considerations. The mechanism of interaction between

lipid membranes and carbohydrates is still in dispute.[106] The effects of alcohols and alkanes have been discussed in Sec. 3.5. Effects of more general anaesthetic compounds, such as cocaine derivatives and halothane, on lipid membrane phase behaviour are described in Ref. 107 where it was found that most anaesthetics increase the lipid chain disorder in fluid membranes. Finally, it should be mentioned that tumor promoting agents, e.g. phorbol esters, apparently also act by modifying membrane structure, lowering the phase transition temperature and fluidizing the membrane.[108] The relative biological potency of different phorbol esters is found to be reflected in magnitude of repulsion between the tumor promoter and the phospholipids.[109]

## 3.8  Physics and the evolution of membranes

The first membranes evolved $3-4 \times 10^9$ years ago from the prebiotic soup by the spontaneous self-organization of lipid molecules into a closed bilayer membrane.[110,111] The first cell was characterized by a lipid bilayer enclosing a selfreplicating RNA-sequence, together with its expressed proteins. The cell membranes at this stage were of the type we now find in procaryotes, i.e. organisms composed by cells without nuclei. The anaerobic procaryotes were dominant up until about $1.5 \times 10^9$ years ago. At that time sufficient molecular oxygen had accumulated in the atmosphere so that a mode of life built upon respiration had to evolve. At this time the eucaryotes come into existence. Eucaryotic cells have a compartmentarized interior, including nucleus and various organelles, each bounded by an internal membrane of its own. This is in brief the standard textbook version of evolution.[110] The question is now: why did it take $10^9$ years for cells to evolve from procaryotes to eucaryotes?

Bloom and Mouritsen[46] have recently proposed an answer to this intriguing question using an argument based on the *physical* properties of membranes. This argument is inspired by the observation that cholesterol appears on the scene at the same time as the first eucaryotes. As proved by Block,[94] cholesterol could not have appeared before, since its biochemical synthesis requires molecular oxygen. Bloom and Mouritsen[46] now propose that cholesterol has removed a bottleneck in the procaryotic-eucaryotic evolution, allowing eucaryotic plasma membranes to evolve. It is a striking observation that cholesterol is universally absent from procaryotic plasma membranes,[112] whereas it is abundant in eucaryotic plasma membranes. It is now suggested[46] that cholesterol, via its unique influence on the mechanical properties of lipid membranes as discussed in Sec. 3.6, makes it possible to form the membranes of the much larger eucaryotic cells and impart the unique capacity of exhibiting cytosis phenomena to these cells, which is a definite characteristic of eucaryotes.[113] A central element of this new evolutionary hypothesis is hydrophobic matching of lipids and proteins, as described in Sects. 3.3-3.5. Introduction of cholesterol releases new evolutionary driving forces by allowing the larger local variations of membrane hydrophobic thickness required for endo- and exocytosis.

## 4. INTERMEMBRANE FORCES

### 4.1 Surface interactions in biology

The organization and functioning of living systems depends crucially upon the long and short range interactions between membranes. This is equally true for inter-actions between different cells of an organism, as well as for processes of transport and communication within the individual cell. Examples of the first type include cell fusion, cytosis via coated vesicles, release of neurotransmitters across synapses via vesicle formation, and stacking of photosynthetic thylakoid membranes. Examples of the second type could be transfer processes of newly synthesized proteins, lipids, and other macromolecules from the organelles of the cell to the plasma membrane, e.g. sec-retion from the Golgi apparatus. It is obvious that, as far as cell fusion is concerned, it is not only a matter of membrane fusion, but the fusion process will also depend on the metabolic activity of the involved cells.[114] Still, it is an important first step to-wards an understanding of this central problem of biology to describe the interaction between lipid bilayers and unravel the different physical forces which are responsible for this interaction.[115]

The basic forces between lipid membranes in an aqueous medium are the van der Waals force, the so-called electric double-layer force, and the hydration force.[116] The ranges of interaction of these forces are illustrated in Fig. 23. The van der Waals

REPULSIVE PRESSURE VS. SEPARATION

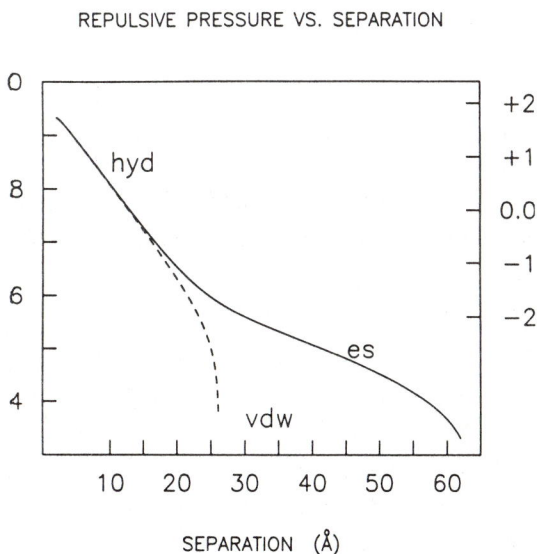

SEPARATION (Å)

Fig. 23. Schematic logarithmic plot showing the strength of van der Waals forces, electric double-layer forces, and hydration forces between lipid bilayers [After Ref. 116].

force is attractive and long-ranged decaying typically as $r^{-3}$ or $r^{-4}$. It depends upon the polarizability of the intermediate medium and is probably too weak to play any important role in membrane fusion.[116] The electrostatic interactions dominate repulsion only at long distances. This leaves the hydration force as the most important barrier to overcome in a fusion process. The hydration force decays exponentially with 20 to 30 Å as a typical correlation range. The hydration force is related to the structure and entropy of the solvent in the neighbourhood of the polar head group region. Hydration forces are not unique for water, but occur also when glycol is used as a solvent.[117] Recently, very ingenious experimental techniques have been developed to measure hydration forces, either by mechanical force balance measurements, or via osmotic pressure studies.[116] It has been found that $Ca^{++}$ ions in very low concentrations have a dramatic effect by eliminating the hydration force and thereby facilitating fusion. Only in much higher concentrations can $Mg^{++}$ ions provide the same effect. This is indeed very intriguing, noting that the $Ca^{++}$ ion balance is essential for proper function at the synaptic junctions in neural networks. The detailed mechanism of $Ca^{++}$ on the membrane surface is still unknown.

A large number of model studies and theoretical calculations have been carried out to describe intermembrane interactions and hydration forces. The reader is referred to Refs. 25, 118 and 119 for some recent important pieces of research. The reader's attention should also be directed to a very recent theoretical development by Lipowsky and Leibler,[26] who by renormalization-group techniques study the interplay between the van der Waals, the electrostatic and the hydration forces. These authors find that there is a continuous fluctuation-driven unbinding transition in bilayer systems (characterized by critical exponents) which separates the stable form of multilamellar aggregates from the state of indefinite swelling.

### 4.2 An Ising model of membrane-membrane interactions

As an example of the use one can make of microscopic interaction models well-known from statistical physics, we shall briefly describe an application of the Ising model formalism to investigate fluctuation-induced forces between lipid bilayers.[120]

It is a characteristic property of membranes (as well as of other colloidal aggregates) that their surfaces contain many internal degrees of freedom and that the interaction between the various polar head groups on the surface could have a strong influence on the effective force between the surfaces. The entropy associated with the distribution and conformation of head groups is essential in order to obtain an effective repulsion. In the limit of a very strong coupling between the head groups of one surface, entropy effects are negligible and there is an attractive interaction between the two surfaces. In the other extreme of a very weak coupling, the head groups of one surface act as a polar, flexible and effective "solvent" for the head groups of the other surface. This also results in an attractive interaction. However, there is an intermediate range where the head groups are disordered relative to one another, but their conformation is not flexible enough to make them a better solvent than the water. In this range a repulsion is obtained. In both cases there is a delicate interplay between energy and entropy effects and between intra- and intersurface interactions. The study of these problems has just begun and one way of obtaining further insight is to study a model system which is simple enough to allow an exact solution.

A convenient version of such a model system consists of two coupled one-dimensional spin - $\frac{1}{2}$ ferromagnetic Ising chains each placed in magnetic fields, B, of opposed direction as given by the Hamiltonian

$$H = -J \sum_{i=1}^{N} (s_i s_{i+1} + t_i t_{i+1}) - B \sum_{i=1}^{N} (s_i - t_i) - K(r) \sum_{i=1}^{N} s_i t_i \ . \tag{7}$$

The spin-$\frac{1}{2}$ variables, $s_i$, $t_i = \pm 1$, imitate the internal degree of freedom on the surface, and the interchain coupling $K(r) \sim r^{-3}$ represents the long-range intersurface interactions. At zero temperature, the sign of the effective total interaction between the chains changes at $B = K$. The partition function for the model can be derived exactly by transfer-matrix techniques.[120] The interesting outcome of the calculation is that at finite temperatures the sign change of the net force no longer occurs at $B = K$, indicating that fluctuations (also beyond the mean-field theory) are important for determining whether the force is attractive or repulsive. A similar situation has been found to arise for charged surfaces in electrolyte systems,[121] where it is the entropy of mixing of the counter-ions that provides the source of the repulsion commonly associated with the overlapping double layers. These results seem to indicate that there is some universality in fluctuation-induced forces in membrane, colloidal and electrolytic systems.

## 5. CONCLUSIONS

Two major lessons taught from the present review are

(a) Membranology presents a number of challenging problems to the physicist,

and
(b) A study of the *physical* properties of biological membranes provides new insight into membrane structure, dynamics, function and evolution.

## Acknowledgements

This work was supored by the Danish Natural Science Council under grant J.nr. 5.21.99.72. The author wishes to express his gratitude to Myer Bloom and Martin Zuckermann with whom he has collaborated on various aspects of the work reviewed in this paper and with whom he has engaged in numerous fruitful discussions on the physics of biological membranes.

## References

1.  B. Alberts, D. Bray, J. Lewis, M. Raff, K. Roberts and J.D. Watson, *Molecular Biology of the Cell*. Garland Publ., New York, 1983. Chap. 6.

2.  S.J. Singer and G. L. Nicolsen, Science 175, 720 (1972).

3.  M.S. Bretscher, Sci.Amer. 253, no. 4, 86 (1985).

4.  R.N. McElhaney, in *Membrane Fluidity in Biology*. (R.C. Aloia and J.M. Boggs, eds.) Academic Press, New York, 1985. p.147.

5.  P.J. Quinn and D. Chapman, CRC Crit.Rev.Biochem. 8, 1 (1980).

6.  J.N. Israelachvili, S. Marčelja and R.G. Horn, Q.Rev.Biophys. 13, 121 (1981).

7.  R.N. McElhaney, Chem.Phys. Lipids 30, 229 (1982).

8.  E.A. Evans and R. Skalak, *Mechanics and Thermodynamics of Biomembranes*. CRC Press Inc., Florida, 1980.

9.  J.L. Ranck, Chem.Phys. Lipids 32, 251 (1983).

10. G. Zaccai, G. Büldt, A. Seelig and J. Seelig, J.Mol.Biol. 134, 693 (1979).

11. P.F. Devaux in *Biological Magnetic Resonance* Vol. 5 (L.J. Berliner and J. Reuben, eds.) Plenum Press, New York, 1983. p.183.

12. J. Seelig and A. Seelig, Q.Rev.Biophys. 13, 1 (1980).

13. M. Shinitzky and I. Yuli, Chem.Phys. Lipids 30, 261 (1982).

14. D.A. Pink, T.J. Green and D. Chapman, Biochemistry 19, 349 (1980).

15. R. Henderson, in *Membranes and Intercellular Communication* (R. Balian, M. Cabre and P.F. Devaux, eds.) North-Holland Publ. Co., Amsterdam, 1981. p.229.

16. N. Unwin and R. Henderson, Sci.Amer. 250, no. 2, 56 (1984).

17. J. Deisenhofer, O. Epp, K. Miki, R. Huber and H. Michel, Nature 318, 618 (1985).

18. S. Marčelja, Biochim.Biophys. Acta 367, 165 (1974).

19. J.F, Nagle, Ann.Rev.Phys.Chem. 31, 157 (1980).

20. A. Caille, D. Pink, F. de Verteuil and M.J. Zuckermann, Can.J.Phys. 58, 581 (1980).

21. P. van der Ploeg and H.J.C. Berendsen, Mol.Phys. 49, 233 (1983).

22. O.G. Mouritsen, *Computer Studies of Phase Transitions and Critical Phenomena*. Springer-Verlag, Heidelberg, 1984. Chap. 5.1.

23. J.R. Abney and J.C. Owicki, in *Progress in Protein-Lipid Interactions* (A. Watts and J.J.H.H.M. De Pont, eds.) Elsevier, Amsterdam, 1985. p.1.

24. D.A. Pink and C.E. Carroll, Phys.Lett. 66A, 157 (1978).

25. B. Jönsson and H. Wennerström, J.Chem.Soc. Faraday Trans. 79, 19 (1983).

26. R. Lipowsky and S. Leibler, Phys.Rev.Lett. 56, 2541 (1986).

27. J.N. Israelachvili, in *Physics of Amphiphilic Micelles, Vesicles and Microemulsions* (V. Degiorgio and M. Corti, eds.) North-Holland, Amsterdam, 1985.

28. B.A. Lewis and D.M. Engelman, J.Mol.Biol. 166, 211 (1983).

29. N. Albon and J.M. Sturtevant, Proc.Natl.Acad.Sci. USA 75, 2258 (1978).

30. J.H. Davis, Biophys.J. 27, 339 (1979).

31. Y. Inoko and T. Mitsui, J.Phys.Soc. Japan 44, 1918 (1978).

32.  E. Evans and R. Kwok, Biochemistry 21, 4874 (1982).

33.  O.G. Mouritsen and M. Bloom, Biophys.J. 46, 141 (1984).

34.  P.R. Cullis, M.J. Hope, B. de Kruijff, A.J. Verkleij and C.P.S. Tilcock, in *Phospholipids and Cellular Regulation* (J.F. Kuo, ed.) CRC Press, Boca Raton, Florida, 1985. p.1.

35.  R.N. McElhaney, Biochim.Biophys. Acta 779, 1 (1984).

36.  J.H. Davis, M. Bloom, K.W. Butler and I.C.P. Smith, Biochim.Biophys. Acta 597, 477 (1980).

37.  G. Durairaj and I. Vijayakumar, Biochim.Biophys. Acta 770, 7 (1984).

38.  R.M. J. Cotterill, Nature 313, 426 (1985).

39.  See e.g. articles in *Membrane Fluidity in Biology* (R.C. Aloia, ed.) Academic Press, New York, 1983. Vol. 2.

40.  E. Lerner, A.L. Shug, C. Elson and E. Shrago, J.Biol.Chem. 247, 1513 (1972).

41.  H. Sandermann, Biochim.Biophys. Acta 515, 209 (1978).

42.  R.N. McElhaney, in *Current Topics in Membranes and Transport* (S. Razin and S. Rottem, eds.) Academic Press, New York, 1982. Vol. 17. p.317.

43.  R.N. McElhaney, in *Membrane Fluidity* (M. Kates and L.A. Manson, eds.) Plenum Publ. Co., New York, 1984. p.249.

44.  M. Sinensky, Proc.Natl.Acad.Sci. USA 71, 522 (1974).

45.  E.F. DeLong and A.A. Yayanos, Science 228, 1101 (1985).

46.  M. Bloom and O.G. Mouritsen, submitted to J.Mol.Evol. 1986.

47.  H. Träuble and D.H. Haynes, Chem.Phys. Lipids 7, 324 (1971).

48.  H. Frischleder and G. Peinel, Chem.Phys. Lipids 30, 121 (1982).

49.  O.G. Mouritsen, D. Boothroyd, R. Harris, N. Jan, T. Lookmann, L. MacDonald, D.A. Pink and M.J. Zuckermann, J.Chem.Phys. 79, 2027 (1983).

50.  O.G. Mouritsen, Biochim.Biophys. Acta 731, 217 (1983).

51.  O.G. Mouritsen and M.J. Zuckermann, Eur.Biophys.J. 12, 75 (1985).

52.  M.J. Zuckermann and O.G. Mouritsen, Eur.Biophys.J. (to be published).

53.  D.A. Pink, Can.J.Phys. 62, 760 (1984).

54.  S. Doniach, J.Chem.Phys. 68, 4912 (1978).

55.  K. Kjaer, J. Als-Nielsen, C.A. Helm, L.A. Laxhuber and H. Möhwald, Phys. Ref.Lett. (in press).

56.  O.G. Mouritsen and M.J. Zuckermann (submitted to Phys.Rev.Lett.)

57.  M.P. Anderson, D.J. Srolovitz, G.S. Grest and P.S. Sahni, Acta Metall. 32, 783 (1984).

58.  B.I. Halperin and D.R. Nelson, Phys.Rev.Lett. 41, 121 (1978).

59.  S. Mitaku, T. Jippo and R. Kataoka, Biophys.J. 42, 137 (1983).

60.  I. Hatta, K. Suzuki and S. Imaizumi, J.Phys.Soc. Japan 52, 2790 (1983).

61.  A. Genz and J.F. Holzwarth, Eur.Biophys.J. 13, 323 (1986).

62.  M. Caffrey, Biochemistry 24, 4826 (1985).

63.  D.L. Melchior, E.P. Bruggemann and J.M. Stein, Biochim.Biophys. Acta 690, 81 (1982).

64.  W. Selke, Surf.Sci. 144, 176 (1984).

65.  J. Kyte and R.F. Doolitte, J.Mol.Biol. 157, 105 (1982).

66. D.M. Engelman, Biophys.J. 37, 187 (1982).

67. See, however, B.K. Jap, M.F. Maestre, S.B. Hayward and R.M. Glaeser, Biophys.J. 43, 81 (1983).

68. T. Ozawa, J. Bioenergetics and Biomembranes 16, 321 (1984).

69. H. Michel, O. Epp and J. Deisenhofer, EMBO J. 5, 2445 (1986).

70. E. Sackmann, in *Biological Membranes*. Academic Press, London, 1984. Vol.5. p.105.

71. For a collection of recent review articles, see *Progress in Protein-Lipid Interactions* (A. Watts and J.J.H.H.M. De Pont, eds.) Elsevier, New York, 1985, 1986. Vols. 1 and 2.

72. D. Chapman, J.C. Gómez-Fernández and F.M. Gõni, Trends.Biochem.Sci., Feb. 1982, p.67.

73. A. Bienvenue, M. Bloom, J.H. Davis and P.F. Devaux, J.Biol.Chem. 257, 3032 (1982).

74. M.R. Paddy, F.W. Dahlquist, J.H. Davis and M. Bloom, Biochemistry 20, 2152 (1981).

75. M.R. Vist, Partial phase behavior of perdeuterated dipalmitoylphosphatidylcholine-cholesterol model membranes. MSc Thesis, University of Guelph, Guelph, Ontario, Canada (1984).

76. F. Jänig, Biophys.J. 36, 329, 347 (1981); J.C. Owicki and H.M. McConnell, Proc.Natl.Acad.Sci. 76, 4750 (1979).

77. S. Marčelja, Biochim.Biophys. Acta 455, 1 (1976).

78. M. Tessier-Lavigne, A. Boothroyd, M.J. Zuckermann and D.A. Pink, J.Chem. Phys. 76, 4587 (1982).

79. J.R. Dahn, D.C. Dahn and R.R. Haering, Solid St.Commun. 42, 179 (1982).

80. J. Riegler and H. Möhwald, Biophys.J. 49, 1111 (1986).

81. D. Chapman, B.A. Cornell, A.W. Eliasz and A. Perry, J.Mol.Biol. 113, 517 (1977).

82. J.H. Davis, D.M. Clare, R.S. Hodges and M. Bloom, Biochemistry 22, 5298 (1983).

83. J.C. Huschilt, R.S. Hodges and J.H. Davis, Biochemistry 24, 1377 (1985).

84. M.R. Morrow, J.C. Huschilt and J.H. Davis, Biochemistry 24, 5396 (1985).

85. J.R. Elliott, D. Needham, J.P. Dilger and D.A. Haydon, Biochim.Biophys. Acta 735, 95 (1983).

86. D.A. Haydon, B.M. Hendry and S.R. Levinsen, Nature 268, 356 (1977).

87. S.H. White, G.I. King and J.E. Cain, Nature 290, 161 (1981).

88. C.L. Stephens and M. Shinitzky, Nature 270, 267 (1977),

89. C. Montecucco, G.A. Smith, F. Dabbeni-sala, A. Johannsson, Y.M. Galante and R. Bisson, FEBS Lett. 144, 145 (1982).

90. A. Johannsson, G.A. Smith and J.C. Metcalfe, Biochim.Biophys. Acta 641, 416 (1981).

91. A. Johannsson, C.A. Keightley, G.A. Smith, C.D. Richards, T.R. Hesketh and J.C. Metcalfe, J.Biol.Chem. 256, 1643 (1981).

92. R.A. Demel and B. de Kruyff, Biochim.Biophys. Acta 457, 109 (1976).

93. P.L. Yeagle, Biochim.Biophys. Acta 822, 267 (1985).

94. K.E. Bloch, CRC Crit.Rev.Biochem. 14, 47 (1983).

95. E. Evans and D. Needham, Faraday Discuss.Chem.Soc. 81 (in press).

96. G. Lindblom, L.B.A. Johansson and G. Arvidson, Biochemistry 20, 2204 (1981).

97. R.J. Cherry, U. Müller, C. Holenstein and M.P. Heyn, Biochim.Biophys, Acta 596, 145 (1980).

98. F.T. Presti, in *Membrane Fluidity in Biology* (R.C. Aloia and J.M. Boggs, eds.) Academic Press, New York, 1985. p.97.

99. D.A. Pink, T.J. Green and D. Chapman, Biochemistry 20, 6692 (1981).

100. C.-H. Huang, Lipids 12, 348 (1977).

101. T.J. O'Leary, Biochim.Biophys. Acta 731, 47 (1983).

102. R.D. Klansner, A.M. Kleinfeld, R.L. Hoover and M.J. Karnovsky, J.Biol.Chem. 255, 1286 (1980).

103. O. Behnke, J. Tranum-Jensen and B. van Deurs, Eur.J.Cell.Biol. 35, 189 (1984).

104. E.J. Dufourc, I.C.P. Smith and H.C. Jarrell, Biochim.Biophys. Acta 778, 435 (1984).

105. A.L.Y. Lau and S.I. Chan, Biochemistry 13, 4942 (1974).

106. B.Z. Chowdhry, G. Lipka and J.M. Sturtevant, Biophys.J. 46, 419 (1984).

107. F. de Verteuil, D.A. Pink, E.B. Vadas and M.J. Zuckermann, Biochim.Biophys. Acta 640, 207 (1981).

108. P.L. Tran, L.T.-M.-Saraga, G. Madelmont and M. Castagna, Biochim.Biophys. Acta 727, 31 (1983).

109. M. Deleers, J.M. Ruysschaert and W.J. Malaisse, Chem.Biol. Interactions 42, 271 (1982).

110. A. Alberts, D. Bray, J. Lewis, M. Raff, K. Roberts and J.D. Watson. *Molecular Biology of the Cell*. Garland Publ. Co., New York, 1983. Chap. 1.

111. R.N. Robertson, *The Lively Membranes*. Cambridge University Press, London, 1983. Chap. 11.

112. T. Cavalier-Smith, in *Molecular and Cellular Aspects of Macrobial Evolution* (M.J. Carlile, J.F. Collins and B.E.B. Mosley, eds.) Cambridge University Press, 1981. p.33.

113. T. Cavalier-Smith, Nature 256, 463 (1975).

114. See e.g. articles in *Cell Fusion* (D. Evered and J. Whelan, eds.) Pitman Books, London, 1984.

115. J. Wilschut and D. Hoekstra, Trends.Biochem.Sci. 9, 479 (1984).

116. V.A. Parsegian, R.P. Rand and D. Gingell, in *Cell Fusion* (D. Evered and J. Whelan, eds.) Pitman Books, London, 1984. p.9.

117. H.K. Christenson and R.G. Horn, J.Coll.Int.Sci. 103, 50 (1985).

118. D.W.R. Gruen and S. Marčelja, J.Chem.Soc. Faraday Trans. 2, 79, 225 (1983).

119. L. Guldbrand, B. Jönsson and H. Wennerström, J.Coll.Int.Sci. 89, 532 (1982).

120. S. Engström, H. Wennerström, O.G. Mouritsen and H.C. Fogedby, Chem.Scr. 25, 92 (1985).

121. L. Guldbrand, B. Jönsson, H. Wennerström and P. Linse, J.Chem.Phys. 80, 2221 (1984).

# TRANSPORT AND SIGNAL TRANSFER ACROSS BIOMEMBRANES

E. Neher

Max-Planck-Institut für biophysikalische Chemie
D-3400 Göttingen, West Germany

## INTRODUCTION

Transport through biomembranes is of importance to living cells in two respects: First of all selective transport of metabolites, ions and other solutes is important for cellular homeostatis in the presence of metabolism. The physical mechanisms underlying this transport include diffusion, facilitated transport through ion channels or carriers, and active transport. Secondly transport, – or better transfer – of signals across membranes is important for cells in order to be able to react to a changing environment or to allow the concentrated action of millions of cells in an organism. This transfer of signals in many cases depends on transport processes through membranes. Best known in this respect are the cation specific ion fluxes of the nerve membrane, which lead to voltage changes across the membrane and to the propagation of the nerve action potential [1]. But transport of matter can also have a different signal character. For instance, the Calcium ion entering a cell, carries a signal apart from its charge. The Calcium concentration inside cells is very low. Therefore even minute quantities of Calcium entering the cell can lead to drastic changes in the concentration of free intracellular Calcium $[Ca]_i$. Many cellular functions, however, like enzymatic activities, muscle contraction, and secretion of hormones and neurotransmitters are critically dependent on $[Ca]_i$. Thus, inflow of Calcium through the membrane can constitute a powerful signal to initiate cellular responses [2].

Apart from these signals, which depend on transport of matter there are also signals crossing the membrane without transport. These signals usually are carried by integral membrane proteins spanning the whole membrane thickness. Upon binding of a ligand to an external receptor they undergo a conformational change which then changes an enzymatic activity of the same molecule localized on the cytoplasmic side. Two major classes of mechanisms of this kind have been elucidated in recent years. In one of these, the molecules in question function as receptors for certain growth factors (EGF, PDGF, gene products of retroviruses) on the exracellular face and express tyrosine-specific protein kinase activity on the cytoplasmic side [3]. The enzymatic activity is switched on and off by the occupancy state of the receptor portion of the molecule. In the other mechanism a receptor can catalyze the exchange of GTP for GDP in special so-called "nucleotide binding proteins" or N-proteins. The GTP-bound form can then activate intracellular enzymes or other effectors. This mechanism, which was first described to hold for activation of adenylate cyclase, has meanwhile been found to be operative in as diverse fields as phototransduction, olfaction and adrenergic stimulation of secretion from glands [4].

This contribution focuses on two well separated aspects of signal transduction in membranes. It first summarizes recent advances in the study of ion channels emphasizing the recording of single channels by the patch clamp method. It then discusses the degree to which ion channels control secretory processes (secretion of neurotransmitters and hormones) and presents some evidence invoking the second kind of signal mechanism – involving GTP-binding proteins – in this signal chain.

## The patch clamp technique for studying single channel currents

Ion channels are membrane proteins, typically of molecular weight 200 – 300 kD, which are believed to undergo conformational changes between different structural states in response to external stimuli. One of these states corresponds to a open pore which can conduct ions across the membrane. The current flow through such a pore under normal physiological conditions is typically in the range pA ($10^{-12}$ A) and channels open typically for time periods of one to several hundred milliseconds. The opening and closing transitions occur stochastically and are fast compared to the lifetimes of open or closed states.

The switching behaviour of single channels can be studied in biological membranes by means of the patch clamp technique [5]. In order to reduce background noise to the degree that signals of the above-mentioned parameters can be resolved, current measurement has to be limited to a very small area of membrane. For this reason a heat-

polished micropipette (diameter 1-3 $\mu$m), filled with physiological
saline, is placed onto the surface of a cell and current flowing
through the pipette is measured. With some precautions regarding
cleanliness of the surfaces involved, and some suction applied to the
pipette an electrically tight seal is obtained between the rim of the
glass pipette and the membrane. Thus, the small area of cell surface
covered by the pipette is singled out for the electrical measurement.
The cell's natural "resting potential" or else voltage applied to the
pipette is the driving force for ion flow through channels which may
happen to open in the small patch of membrane under investigation.
Typical recordings of single channel currents activated by the
neurotransmitter Acetylcholine (ACh) are shown in figure 1. In part A
the concentration of ACh is very low, such that opening of a channel is
a rare event. In part B the concentration is high such that a single
channel fluctuates stochastically between open and closed states.

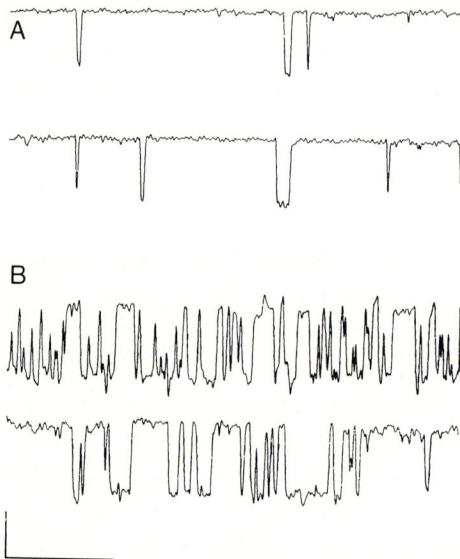

Fig. 1 – Recordings of membrane current from a bovine chromaffin cell
exposed to 2 $\mu$M acetylcholine (part A) and 20 $\mu$M acetylcholine
(part B); two traces each. Membrane voltage was -80 mV. A
downward deflection represents increase in inwardly directed
current. The current pulses in part A probably originate from
many channels. In part B conditions were chosen such that only
one channel is operating during he record (see [18] for
further detail). Experiments were performed in collaboration
with Dr. D.E. Clapham. Reproduced with permission from [19].

Once a seal has been obtained between glass pipette and membrane, the patch sticks very tightly to the glass such that microscopic membrane patches can be excised from the biological preparation and can be studied in isolation. Also, the patch can be broken by electrically discharge or by a pulse of suction leading to a conducting pathway between measuring pipette and the interior of the cell under study. This allows investigation of "whole-cell" membrane currents [6].

## Many types of ion channels carry a wide range of signals in the nervous system

In 1952 Hodgkin & Huxley [1] published their theory of nerve excitability in which they identified the upstroke of the nerve impulse with an inflow of $Na^+$-ions into the cell. The downstroke or repolarization is due to a delayed outflow of $K^+$-ions. These ion movements are mediated by two types of voltage gated ion channels, one being selective for Na-ions, the other one for $K^+$-ions. They have been identified on the single-channel level, and together convey to nerve fibers the property to generate and propagate action potentials. However, a nerve cell has to fulfil many more tasks. It has to receive incoming signals from other neurones or peripheral receptors, it has to integrate and process these signals, and it has to liberate neurotransmitter in order to convey its outgoing signal to subsequent neurones or to effector organs. In a typical synapse, (the contact region between two neurones), a presynaptic nerve ending releases neurotransmitters in response to a nerve action potential. This release process depends on at least two types of ion channels (see below), a Calcium selective channel, and a special $K^+$-selective channel. The neurotransmitter acts on the postsynaptic membrane (the membrane of the receiving cell) by opening ion channels. Since different cells store many different transmitter substances there are also many (probably 10 to 20) different ion channels with receptors specific for certain transmitters [7]. Activation of these channels can either depolarize or hyperpolarize the receiving cells, depending on the ion selectivity of the channels and on the relevant ionic gradients.

The cell's response to such stimuli can be quite variable, depending on another set of ion channels, so-called modulated channels, the properties of which can be slowly changed (modulated) by other transmitters, neuropeptides or hormones. Recent research has shown that also some of the classical channels like Ca-specific channels can be modulated. Very often modulation involves a phosphorylation-dephosphorylation reaction which can profoundly change the gating properties of a channel. Also, modulation can be exerted by the GTP-binding proteins mentioned in the introduction [9].

## Control of secretion

The classical concept of secretion control derives from studies on the neuromuscular junction, where it was shown by Katz & Miledi [10] that the release of neurotransmitters from the nerve ending is a consequence of the inflow of $Ca^{++}$. The nerve ending carries voltage dependent Ca-specific ion channels which open during the action potential. This leads to a marked increase in the concentration of free intracellular Calcium, $[Ca]_i$, since this parameter is always kept very low in resting cells. Increased $[Ca]_i$ somehow causes the fusion of transmitter-storing granules with the outer membrane, thereby releasing the granule contents into the extracellular space.

This concept of Ca-mediated release control, which certainly is important in neurotransmitter release, was believed to be generally applicable to basically all regulated secretory processes, such as hormone release from endocrine glands, thrombin release from blood platelets, histamine release from mast cells and enzyme release from exocrine glands. However, recent advances in Ca-measurement techniques [11] have shown that in some cell types secretion can proceed without any elevation of $[Ca]_i$. At the same time it has become clear that in most of these cell types a very prominent intracellular signalling mechanism can be identified, relying on the enzymatic splitting of the membrane lipid phosphatidylinositol [12]. Already in 1953 Hokin & Hokin have shown that hormonal stimulation of pancreas led to increased metabolic turnover in phospholipids. It is now established that for instance in mast cells an external stimulus leads to the activation of a phospholipid splitting enzyme [13], the phosphodiesterase (PDE) or phospholipase C (see fig. 2). This coupling between the external receptors and the PDE is mediated by a GTP-binding protein as mentioned in the introduction. The two breakdown products of this reaction have second messenger functions. One, the water-soluble inositol-triphosphate ($IP_3$) can release Calcium from intracellular stores and thus contribute to secretion. The other one - membrane-bound

# Stimulus-Secretion Coupling in Neurones

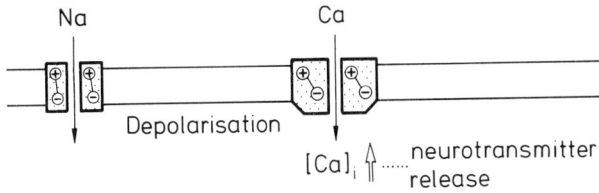

# Models Proposed for Mast Cells:

A) Channels

B) PI breakdown

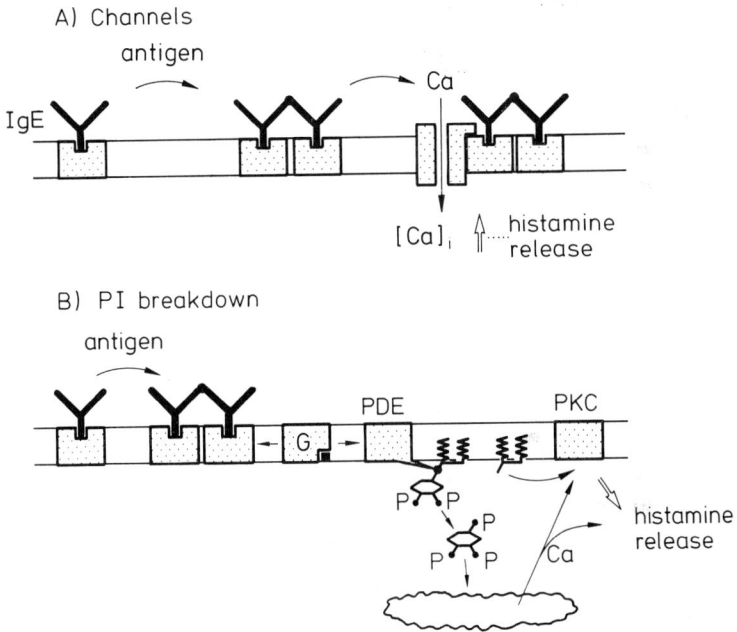

Fig. 2 – Various models for secretion in neurones and mast cells (see text for explanation). Abbreviations: IgE: Immuno-globulin E present on the surface and bound to specific IgE-receptors of sensitized mast cells; G: GTP-binding protein; PDE: phosphodiesterase (phospholipase C) which splits phosphatidylinositol; PKC: protein kinase C. The closed compartment at the bottom is supposed to represent the endoplasmic reticulum.

diacylglycerol - can associate with and thereby activate protein kinase C [12]. This, in turn, leads to various cellular responses. There is increasing evidence that secretion in mast cells and in a number of other cell types is mediated by this pathway. Alternatively, the traditional view of secretion in mast cells proposes that the initial stimulus (which is dimerization of antigen receptors) leads to the opening of Ca-channels in analogy to the case in neurones (see fig. 2).

We (E. Neher in collaboration with A. Marty, J. Fernandez, M. Lindau, B. Gomperts and W. Almers) have combined patch clamp techniques with a method to measure $[Ca]_i$ in single cells in order to be able to resolve some of these questions by studying simultaneously membrane currents, $[Ca]_i$, and the secretory process. First, we developed a method which would allow to monitor secretion electrically [14,15]. This involves measurement of membrane capacitance which increases during secretion as the membrane of storage vesicles is incorporated into the cell's outer membrane. Then, we employed the new fluorescent Ca-indicator dye fura-2 with which changes of $[Ca]_i$ in the range 0.1-1 $\mu M$ can be readily measured in single cells, using a two wavelength technique [16].

Our results so far are all in line with the PI-breakdown model. On the other hand we did not succeed to identify Ca-channels in mast cells as postulated by the channel model [17]. It should be noted that in neurone-like cells like chromaffin cells Ca-currents as large as several hundred pA can readily be recorded; any similar currents in mast cells which might have escaped our attention should be smaller than a few picoamperes, however.

In contrast findings which are in favour of the PI-breakdown model are:

1. GTP-$\gamma$-S, a nonhydrolyzable analogue of GTP reproducibly activates mast cells when introduced intracellularly. This probably reflects permanent activation of the GTP-binding protein.

2. Both GTP-$\gamma$-S and $IP_3$ can induce transient increases of $[Ca]_i$ which are not dependent on extracellular Calcium and thus should result from release of Calcium from intracellular stores.

3. The relation between $[Ca]_i$ and secretion was found not to be straightforward. An increase in $[Ca]_i$ was not a sufficient stimulus for secretion. This may reflect the fact that in the PI-breakdown model two signals have to converge in order to produce secretion.

In contrast, in neurone-like cells we did find Ca-currents and a clear correlation between $[Ca]_i$ and the rate of secretion. The question remains why nature uses different mechanisms in different cell types to control similar processes. The answer may lie in specific requirements of neurones which have to be able to react fast. Possibly a multistep enzymatic mechanism would not be able to produce a response in the time range milliseconds. Thus neurones had to develop a fast release mechanism which depends on rapid changes in submembrane $[Ca]_i$ when Ca-channels open. At the same time this mechanism would establish the link to the fast electrical signals in neurones which are based on ion channels.

## REFERENCES

[1]  A.L. Hodgkin and A.F. Huxley, "A quantitative description of membrane current and its application to conduction and excitation in nerve", J. Physiol. (Lond.) 297, 1-8 (1952).

[2]  H. Rasmussen, "The Calcium messenger system", New England Journal of Medicine 314, 1094-1101 (1986).

[3]  T. Hunter and J.A. Cooper, "Protein-tyrosine kinases", Ann. Rev. Biochem. 54, 897-930 (1985).

[4]  H.R. Bourne, "One molecular machine can transduce diverse signals", Nature 321, 814-816 (1986).

[5]  O.P. Hamill, A. Marty, E. Neher, B. Sakmann, and F.J. Sigworth, "Improved patch-clamp techniques for high-resolu-tion current recording from cells and cell-free membrane patches", Pflügers Arch. 391, 85-100 (1981).

[6]  A. Marty and E. Neher, "Tight-seal whole-cell recording", in Single Channel Recording, B. Sakmann and E. Neher, eds. (Plenum Press, New York, 1983), pp. 107-122.

[7]  P. Adams, "Transmitter-evoked channels in mammalian central neurons", Trends in Neurosci. 7, 135-137 (1984).

[8]  W. Osterrieder, G. Brun, J. Hescheler, W. Trautwein, V. Flockerzi, and F. Hofmann, "Injection of subunits of cyclic AMP-dependent protein kinase into cardiac myocytes modulates Ca2+ current", Nature 298, 576-578 (1982).

[9]  G.E. Breitwieser and G. Szabo, "Uncoupling of cardias muscarinic and $\beta$-adrenergic receptors from ion channels by a guanine nucleotide analogue", Nature 317, 538-540 (1985).

[10] B. Katz, The release of neural transmitter substances, (Liverpool University Press, Liverpool, 1969).

[11] R.Y. Tsien, T. Pozzan and, T.J. Rink, "Measuring and manipulating cytosolic Ca with trapped indicators", Trends in Biochem. Sci. 9, 263-266 (1984).

[12] M.J. Berridge, "Inositol triphosphate and diacylglycerol as second messengers", Biochem J. <u>220</u>, 345-360 (1984).

[13] B.D. Gomperts, "Calcium shares the limelight in stimulus-secretion coupling", Trends in Biochem. Sci. <u>11</u>, 290-292 (1986).

[14] E. Neher and A. Marty, "Discrete changes of cell membrane capacitance observed under conditions of enhanced secretion in bovine adrenal chromaffin cells", Proc. Natl. Acad. Sci. USA <u>79</u>, 6712-6716 (1982).

[15] J.M. Fernandez, E. Neher, and B.D. Gomperts, "Capacitance mesurements reveal stepwise fusion events in degranulating mast cells", Nature <u>312</u>, 453-455 (1984).

[16] E. Neher and W. Almers, "Fast calcium transients in rat peritoneal mast cells are not sufficient to trigger exocytosis", EMBO J. <u>5</u>, 51-53 (1986).

[17] M. Lindau and J.M. Fernandez, "IgE-mediated degranulation of mast cells does not require opening of ion channels", Nature <u>319</u>, 150-153 (1986).

[18] B. Sakmann, J. Patlak, and E. Neher, "Single acetylcholine-activated channels show burst-kinetics in presence of desen-sitizing concentrations of agonist", Nature <u>286</u>, 71-73 (1980).

[19] E. Neher, "Ion flow through individual membrane pores and its control by membrane voltage and by chemical reactions", in: *Chemistry for the Future*, H. Grünewald, ed. (Pergamon Press, Oxford, 1984), pp. 275-280.

CHEMICAL INSTABILITIES AND APPLICATIONS
OF BIOLOGICAL INTEREST

John Ross
Department of Chemistry
Stanford University
Stanford, California 94305

Chemical reaction systems with complex reaction mechanisms, sustained far from chemical equilibrium display a variety of interesting processes, collectively sometimes
called chemical instabilities or non-linear chemical phenomena.[1,2] We give a brief
summary of some of these processes and cite applications of possible biological interest.

## 1. Multiple Stationary States

There are many chemical systems which, far from equilibrium, have multiple stationary
states. An example is the reaction $S_2O_6F_2 = 2SO_3F$ irradiated by light with wavelength chosen to be absorbed by $SO_3F$ only.[3] Light is absorbed and turned into heat;
the temperature increases and the monomer concentration is thereby increased. This
positive feedback loop leads to multiple stationary states: on increasing the light
intensity and subsequently decreasing it, chemical hysteresis is predicted[4] and observed[3]. The unstable branch of stationary states can be stabilized by means of an
external feedback loop which does not affect the location of those states in the
phase space of the variables (concentration and temperature) ; this procedure allows
the experimental determination of the unstable branch, relaxation studies, and an
estimate of the influence of fluctuations.

A chemical system with multiple stationary states is a chemical switch and the possible use of that concept in a biological context has been proposed.[6] Some biological
species exist in more than one distinct state of metabolic activity and models with
multiple stationary states have been suggested.

## 2. Chemical Oscillations

Concentrations of chemical intermediates and products may vary in time regularly for
complex reaction mechanisms, with auto or cross catalysis or other feedback loops,
run far from equilibrium. Many chemical reactions, in the gaseous[7,8] or liquid[9] phase
are known, and a review of biological oscillatory reactions lists[10] about 150 reactions, including glycolysis, fluxes across mitochondria, neuronal activities, cell
agregation (slime molds), and many other types. Many of these oscillatory reactions
are so-called relaxation oscillation limit cycles,[11] with stable amplitude and

frequency, marginally stable phase, and multiple time scales.  A possible role of oscillatory reactions in biological systems is cited in the last section.

3.  Chemical Fronts, Pulses, Waves.

Travelling concentration gradients in a chemical system are called fronts, pulses, and waves, depending on the type of travelling structure.[12]  Many observations of such temporal and spatial structures have been reported.  Absorption[13] and photo-graphic[14,15] measurements of these concentration gradients have been made.  The am-plitude and shape of the travelling concentration profiles are constant in time[13] in excitable chemical systems (reaction mechanisms which can oscillate but are main-tained at a stationary state close to a bifurcation to oscillations), in close sim-ilarity to nerve conduction signals.[16]

4.  Externally Perturbed Reactions

Consider an oscillatory reaction subjected to external periodic perturbations of arbitrary magnitude of one or more of the chemical species or one or more of con-straints, such as the influx of reactants in an open system.  The perturbation, if sinusoidal for example, is characterized by an amplitude and frequency.  The re-sponse of the systems, dependent on these quantities, may be periodic (entrained), quasi-periodic, or chaotic.[17,18,19]  Experiments on both single and two perturba-tions[18] have been made; the temporal responses and their phase relations to the perturbations and to each other, are likely valuable sources of information on these complex reaction mechanisms.

External perturbations of non-linear reaction mechanisms not under oscillatory con-ditions may also lead to interesting effects, as discussed in the next section.

5.  Efficiency of Biological Reactions: Proton Pumps and Glycolysis

A chemical reaction run in a stationary state with concentrations of reactants and products fixed has a given Gibbs free energy change, $\Delta G$, and a given dissipation (which, at constant temperature, is the product of $\Delta G$ times the rate of the reac-tion.[10])  The imposition of an external periodic perturbation on a reaction in a stationary state effects both an oscillatory $\Delta G$ and an oscillatory rate.  For cer-tain non-linearities in the reaction mechanism the external perturbation can further effect phase shifts of $\Delta G$ and the rate which depends on the frequency (and possibly, but likely less so, on the amplitude) of the perturbation.  The dissipation thus varies from that at a stationary state dependent on the frequency of perturbation. With a decrease in dissipation the efficiency of Gibbs free energy in the reaction transfer is increased.

Theoretical studies have been made of the efficiency of a proton pump[20] and its

variation (both increases and decreases) with temporal periodic variation of the concentration of ATP (the energy input into the pump and utilization of ATP) as well as
the efficiency of glycolysis[21-24], that is the production of ATP, which is autono-
mously oscillatory[25] under constant glucose consumption.  Both processes can be more
efficient under oscillatory than under stationary state conditions, and this increase
in efficiency has been cited as a possible reason for the incorporation of oscillatory reactions in the evolutionary development in biological systems.[21-24]

## References

1.  Synergetics An Introduction, by H. Haken, Springer, 2nd Ed. 1978;
    Synergetics A Workshop, Editor H. Haken, Springer 1977;
    Synergetics Far from Equilibrium - Editors: A. Pacault, C. Vidal, Springer 1979;
    Pattern Formation by Dynamic Systems and Pattern Recognition, Editor: H. Haken,
        Springer 1979;
    Dynamics of Synergetic Systems, Editor: H. Haken, Springer, 1980;
    Stochastic Nonlinear Systems in Physics, Chemistry, and Biology
        Editors: L. Arnold, R. Lefever, Springer, 1981;
    Chaos and Order in Nature, Editor: H. Haken, Springer, 1981;
    Nonlinear Phenomena in Chemical Dynamics, Editors: C. Vidal, A. Pacault, Springer,
        1981.

2.  Noise-Induced Transitions Theory and Applications in Physics, Chemistry and
        Biology, by W. Horsthemke, R. Lefever, Springer 1984;
    Physics of Bioenergetic Processes, by L. A. Blumenfeld, Springer 1983;
    Chemical Oscillations, Waves, and Turbulence, by Y. Kuramoto, Springer 1984;
    Advanced Synergetics, by H. Haken, Springer 1983.

3.  C.L. Creel and John Ross, J. Chem. Phys. 65, 3779 (1976);
    E. C. Zimmermann and John Ross, J. Chem. Phys. 80, 720 (1984);
    Jesse Kramer and John Ross, J. Phys. Chem. 90, 923 (1986).

4.  A. Nitzan and John Ross, J. Chem. Phys. 59, 241 (1973).

5.  E. C. Zimmermann, Mark Schell, and John Ross, J. Chem. Phys. 81, 1327 (1984).

6.  B. Hess, A. Boileaux, and E. E. Selkov, Hoppe-Seylers Physio. Chemie. 361, 610
        (1980).

7.  Second European Symposium on Combustion, P. G. Felton, B. F. Gray and N. Shank
    (The Combustion Institute, Orleans, France, 1975).

8.  B. F. Gray and J. C. Jones, Combust. Flame 57, 3 (1984).

9.  Oscillations and Travelling Waves in Chemical Systems, R. Field and M. Burger,
    Wiley, New York, 1985.

10. P. E. Rapp, J. Exp. Bio. 81, 281 (1979).

11. C. Vidal and A. Pacault, in Evolution of Order and Chaos, Edited by H. Haken,
        Springer-Verlag, Heidelberg, 1982).

12. C. Vidal and P. Hanusse, Int. Rev. Phys. Chem. 5, 1, 55 (1986);
    H. Eyring and D. Henderson, Theoretical Chemistry 4, Periodicities in Chemistry
        and Biology, Academic Press, New York, 1978.

13. Peggy Marie Wood and John Ross, J. Chem. Phys. 82, 1924 (1985); J. M. Bodet, John Ross and C. Vidal - in press

14. S.C. Müeller, T. Plesser and B. Hess, Naturwissensch 73, 165 (1986).

15. A. Pagola and C. Vidal, J. Phys. Chem. - in press

16. R. Fitz Hugh, Biophysics J., 1, 445 (1961).

17. F. W. Schneider, Periodic Perturbations of Chemical Oscillators: Experiments, Annual Rev. of Phys. Chem. 36, 347 (1985).

18. Spencer A. Pugh, Mark Schell and John Ross, J. Chem. Phys. 85, 868 (1986); Spencer A. Pugh, Bruce DeKock and John Ross, ibid, 85, 879 (1986).

19. M. Markus, D. Kuschwitz and B. Hess, FEBS, 172, 235 (1984).

20. Mark Schell, Kaylan Kundu and John Ross, PNAS-U.S. (1987). In press.

21. Yves Termonia and John Ross, J. Chem. Phys. 74, 2339 (1981).

22. Yves Termonia and John Ross, PNAS-U.S., 78, 2952 (1981).

23. Yves Termonia and John Ross, PNAS-U.S. 78, 3563 (1981).

24. Yves Termonia and John Ross, PNAS-U.S. 79, 2878 (1982).

25. Biological and Biochemical Oscillators, B. Chance, R. K. Pye, A.M. Gosh and B. Hess eds, Academic Press, New York, 1973; A. Goldbeter, S. R. Caplan, Annual Rev. Biophys. Bioeng. 5, 449 (1976).

# THE INNERVATION OF SKELETAL MUSCLES: PROPERTIES EMERGING FROM A RANDOM NEURAL NETWORK

Hans-R. Lüscher, Department of Physiology, University of Bern
Bühlplatz 5
CH - 3012 BERN, Switzerland

## Introduction

Even relatively simple movements involve the coordination of many separate muscles which may act at different joints with several degrees of freedom. To move a limb precisely in time and space, requires a set of regulatory mechanisms for the muscles to produce the necessary tensions and lengths. In this lecture I would like to dwell on the question of how a single muscle is controlled to produce the force output adequate for a particular task.

The motor unit: The nerve cells called motoneurons, located in the ventral horn of the spinal cord exert control over a particular muscle. A single motoneuron innervates a certain number of muscle fibres. A nerve impulse, set up in one motoneuron is conducted down its nerve fibre and into all its branches to the neuromuscular junctions, where it initiates impulses in all the muscle fibres leading to their synchronous contraction. This functional entity is called a motor unit. The size of the motoneuron is matched to the size of its axon and the number of muscle cells it innervates. Thus, a small motoneuron innervates a small number of muscle fibres, whereas a larger motoneuron supplies a larger number of muscle fibres. Therefore, the twitch tension a motor unit can produce is proportional to its size.

The motoneuron pool and the Size Principle: The group of motoneurons innervating a particular muscle is called a pool. The motoneuron pool supplying the medial gastrocnemius muscle of the cat, consists of about 300 motoneurons of different sizes. This large number of motor units with a wide range of contractile properties are at the disposal of the central nervous system. How should it select those it needs for a particular task and how should it combine them to produce the muscle tension it desires? It is conceivable that the central nervous system could select any combination of cells to yield the proper total.

This implicitly assumes that the central nervous system is capable of addressing each motoneuron separately. It would also necessitate neural circuitries to calculate the best combination of a subset of the pool to produce the correct total tension. For a pool of 300 motoneurons the number of possible combinations is greater than $10^{90}$. The central nervous system is not capable to perform the necessary calculations in time and no separate input channel to each motoneuron could be found. The solution that has evolved is the so called rank-ordered pool, which provides a simple rule for the appropriate combination and activation of the motor units.

The motoneuron pool is a hierarchy of cells, organized according to their sizes. The rank order of the cells in this hierarchy is defined by their excitability to produce an action potential. The lowest ranking, smallest motoneuron has the highest excitability, whereas the highest ranking cell, the largest motoneuron, is the least susceptible to discharge. Whenever a particular motoneuron gets activated, all lower ranking cells within the pool are activated with it. This simple law of combination of motor units emerged from experimental data and has been described by Henneman and his collaborators in a classical series of papers published in 1965 (Henneman et al. 1965a, b).

In more physiological terms, during a smoothly graded contraction of a skeletal muscle, the component motor units are brought into activity in order of their sizes. This observation has become known as the "Size Principle" of recruitment order of motor units. The smallest motor units are invariably recruited first in any contraction of a skeletal muscle. More force is produced by adding, in an incremental manner the next larger motor unit. This rank order in the recruitment of motor units is invariant and independent of the contraction speed of the muscle. It can be shown, that the isolated spinal cord, together with excitatory input from peripheral sensory receptors possess the ability to produce the size related recruitment order. This stable, ordered function must therefore emerge from the neural network formed by the afferent excitatory input and the motoneuron pool. In what follows I would like to trace back the emergence of the Size Principle to the intimate interrelation of function and structure in a random complex neural network.

The cellular basis of the Size Principle: The cellular basis of the Size Principle in the recruitment order is illustrated in fig. 1. The motoneurons are represented in an oversimplified way as spherical cells. Four cells of different sizes should stand for a population of about 300 motoneurons. An homogeneous excitatory synaptic input of a given strength produces a suprathreshold excitatory synaptic potential (e.p.s.p.) in the smallest motoneuron, so that cell fires an action potential. The same input produces progressively smaller e.p.s.p.s in successively larger cells. If the excitatory drive becomes greater, the next larger motoneuron becomes active and is recruited.

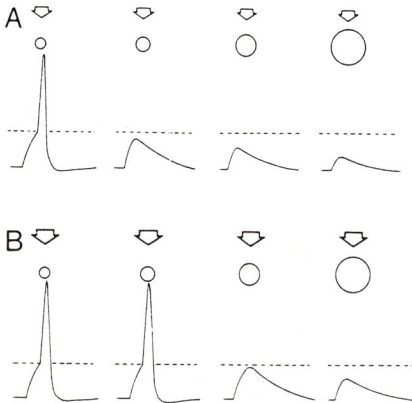

Fig. 1

Cellular basis for a size principle of motoneuron recruitment. A homogenous excitatory input (arrows) to four motoneurons of different sizes produces e.p.s.p.s with amplitude inversely related to cell size. A: The e.p.s.p. in the smallest cell reaches threshold to give an action potential. B: If the excitatory input increases, the second smallest cell also fires an action potential. (after Lüscher, 1983)

In fact it has been shown experimentally by several authors that the amplitude of the e.p.s.p. evoked by synchronous stimulation of the homonymous muscle afferent fibres are inversely proportional to the size of the motoneuron they are recorded from (Burke 1968, Lüscher et al. 1979). Clearly, this observation does not explain the Size Principle, it merely shifts the operation of the principle, first observed peripherally at the innervation pattern of skeletal muscle, to the level of the motoneuron.

Some simple biophysical aspects of the e.p.s.p.: During synaptic transmission, a chemical substance is released from the presynaptic nerve endings. This substance, called transmitter, increases the conductance for cations at the postsynaptic membrane just opposite to the presynaptic ending, leading to an inward current which exits the cell through the adjacent membrane areas. This synaptic current (Is) leads to a transient depolarization of the resting membrane potential. This depolarization represents the excitatory postsynaptic potential. Fig. 2 illustrates in a simplified manner, how the input resistance of the cell and the number of terminals given off to a single cell determines the amplitude of the e.p.s.p. Provided that the synaptic current is similar in any motoneuron, a single synaptic terminal will produce a larger e.p.s.p. in a small cell than in a large cell, simply because the voltage drop is larger across the higher input resistance of the small cell.

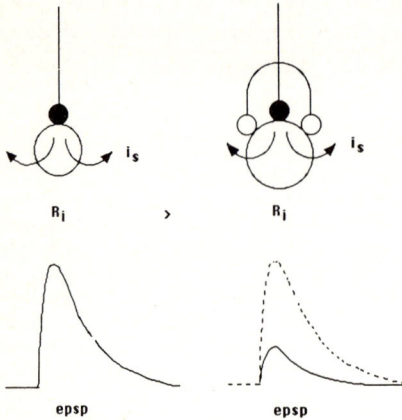

Fig. 2

Scheme to illustrate how the number of synaptic terminals and the input resistance of the motoneuron determine the amplitude of the e.p.s.p.

This assumption would suffice to explain the Size Principle if not the number of terminals on large cells would be proportionally larger. Thus, the higher number of terminals on large cells would compensate for the lower input resistance, leading to an equal size of the e.p.s.p. in small and large cells. There seems to be a discrepancy between the biophysical requirements for a Size Principle and the anatomical suggestion of an equal density of synaptic endings on small and large motoneurons (Burke, 1968; Zucker, 1973; Lüscher, 1981).

A simplified view on a sensory-motor control system: Figure 3 merely serves to give a brief orientation for the sensory motor system which I would like to discuss in more detail later. Stretch of the medial gastrocnemius muscle is sensed by specialized organs, called muscle spindles. About 60 spindles convey the information of stretch, by means of trains of action potentials carried in the afferent fibres, to the 300 motoneurons located in the ventral horn of the spinal cord.

The motoneurons get excited through synaptic transmission. Increasing stretch to the muscle increases the excitatory drive to the motoneuron pool. As a result more and more motoneurons get recruited according to their sizes. The muscle contracts opposing the externally applied stretch. This is called a stretch reflex and has been the object of our experimental studies since many years.

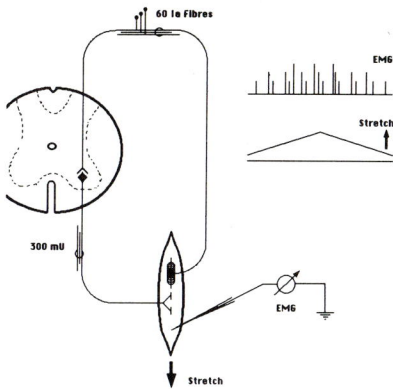

Fig. 3

Scheme of the tonic stretch reflex. Further details in text.

## The morphological bases of the central neural network

Before discussing the functional properties of the central neural network responsible for the stretch reflex, some of its morphological
aspects should be illustrated. This work stems mainly from studies by
Burke (Burke et al. 1978) and Brown (Brown et al., 1978) and their
collaborators (for further references see Brown, 1981). Nerve cells
with all their processes and ramifications can be made visible by intracellular application of the enzyme horse radish peroxidase (HRP).
This enzyme, after suitable histochemical processing, forms a dark reaction product which allows microscopic reconstruction of the entire
nerve cell. As we have seen, the stretch reflex is mediated by two
classes of neurons which differ fundamentally in their structure.

The motoneuron: The first neural elements of the network are the motoneurons. They are by no means spherical cells. Fig. 4 shows that a motoneuron possesses a small cell body some 30-70 $\mu$m in diameter from
which spreads a complex dendritic apparatus. The function of these
dendrites is to capture synaptic contacts from nerve fibres which are
growing towards them during development. This dendritic meshwork pe-
netrates a roughly spherical volume of neural tissue with a diameter
of about   2 mm. The question of large and small cells is no longer
easy to answer. The spatial extent of the cell does not appear to be
significant. Rather it is the density of the dendritic apparatus penetrating through a given volume of neuropile and the surface area available for establishing synaptic contact with ingrowing afferent nerve
fibres that are important (Westbury, 1982).

Thus, the cell illustrated in fig. 4A. corresponds to a large motoneuron, while a small cell is shown in fig. 4C. The structural plan of the motoneuron is the same for all cells, but each motoneuron is unique in the details of its structure.

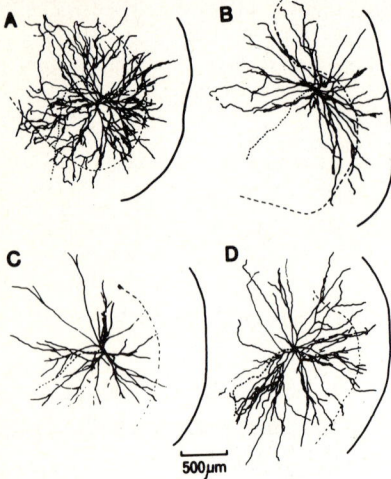

Fig. 4

Reconstructions of four α-moto-neurons innervating the triceps surae (after Brown and Fyffe 1981).

Since the cell bodies of the motoneurons are packed quite densely, the intermingled dendrites of different cells form a intricate meshwork.  Interwoven into this felt of dendritic processes are now the second elements of the neural network, the afferent fibres.

The afferent input: Figure 5 shows how the nerve fibre from a muscle spindle primary ending bifurcates into an ascending and descending branche after entering the spinal cord. At regular intervals these branches give off collaterals which penetrate the spinal cord in a dorsoventral direction. These collaterals, enlarged in the inset, form elaborate arborizations. Because of the intimate intermingling of the afferent fibre with the dendritic meshwork of the motoneurons, each afferent fibre is capable of contacting each motoneuron. The number of individual synaptic contacts formed and their spatial arrangement on the dendritic tree is a direct result of the local structural features of the neural elements involved. Therefore, it would be impossible to discover the expected relation between motoneuron size and numbers of synapses by morphological analysis of individual contact systems. A single contact system, one example is shown in the enlarged inset, can include 3 to 40 individual boutons. It must be noted here, that the synapses from the afferent fibres we are discussing, contribute

probably less than 5% to the total synaptic population of the motoneuron pool.

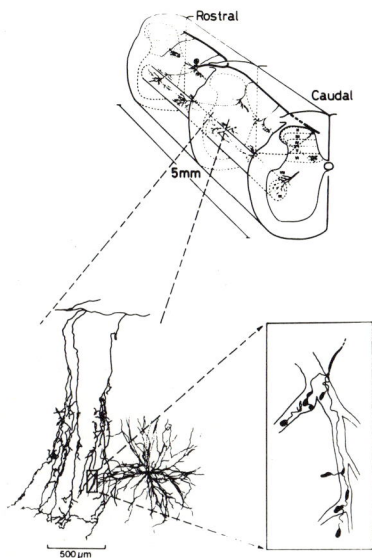

Fig. 5

A: Organization of the Ia fibre
in the spinal cord. B: Enlarged
portion of the collateral fibres forming a complex terminal
arborization. C: Two examples
of single synaptic contact systems (after Brown 1981).

Is the network random ? Given the great structural individuality of
each motoneuron and each terminal arborization of the afferent fibre,
the complex neural network seems to be formed at random. It suggests
that synapses are established at every possible change encounter between an afferent fibre and a motoneuron's dendrite. We will later see
how size related structural constraints place limits on the probabilistic formation of this network.

The neural network described above represents the hardware for a
functional network responsible for the size related recruitment order
during the stretch reflex. As we will see, functional connectivity
does not parallel necessarily anatomical connectivity as given by the
hard wired network.


Functional connectivity versus anatomical connectivity

The whole network consists of roughly 18'000 synaptic contact systems
(300 motoneurons and 60 Ia-fibres), each consisting of 4 synaptic boutons on average. Probably many more synaptic contact systems are formed by spindle group II fibres. We have developed a technique which
enables us to study the functional properties of up to 260 synaptic
contact systems in a single experiment (Lüscher et al., 1983). It is

the largest number of synapses ever studied simultaneously in a defined neural network of a vertebrate nervous system, but nevertheless it only represents probably less than 1.0% of the total. The technique makes it possible to record the synaptic potential evoked by a single afferent action potential in a single motoneuron. Such an e.p.s.p. is called an individual e.p.s.p. Because it is possible to record and distinguish the action potentials in many different afferent fibres simultaneously, the different individual e.p.s.p. they evoke in a single motoneuron can be analysed. Furthermore it is also possible to record the individual e.p.s.p. evoked by the same set of afferent impulses in many motoneurons. This enables us to reconstruct the functional properties of a subset of the network and to relate them to structural properties of the same subset.

The single fibre e.p.s.p.: Two sets of individual e.p.s.p.s recorded from two different motoneurons in the same experiment are reproduced in fig. 6. Those on the left side were recorded from a large motoneuron with an axonal conduction velocity of 91 m/sec., while those on the right side were recorded from a small motoneuron with an axonal conduction velocity of 75 m/sec. The e.p.s.p.s in the two columns were elicited by action potentials in the same 11 afferent fibres. The axonal conduction velocities of the afferent fibres, as a measure of their sizes, are given between the tow columns of responses. They are arranged in descending order from 94 to 32 m/sec. Each e.p.s.p. represents the average of 1024 responses. A multiplicity of sizes and shapes of the individual e.p.s.p.s is evident. This probably reflects directly the great variability in the structural details of the single synaptic contact system.

The amplitude of the individual e.p.s.p. is not clearly related to the size of the motoneuron as one would expect from the studies on e.p.s.p.s elicited by stimulating large numbers of afferent fibres simultaneously. However, what can be assumed from this illustration and what has clearly been shown in the ensemble of many such experiments is, that only large afferent fibres have the capacity to elicit large e.p.s.p.s but they can as well produce small ones. In each of the two columns, one afferent fibre does not produce any e.p.s.p. Failure to produce a functional connection does not always imply that no anatomical connection may exist, since by suitable means or spontaneously inactive connections can be made functional.

If we regard connectivity as an all-or-none response, disregarding the amplitude of the individual e.p.s.p., a connectivity matrix of the connections studied in a single experiment can be compiled.

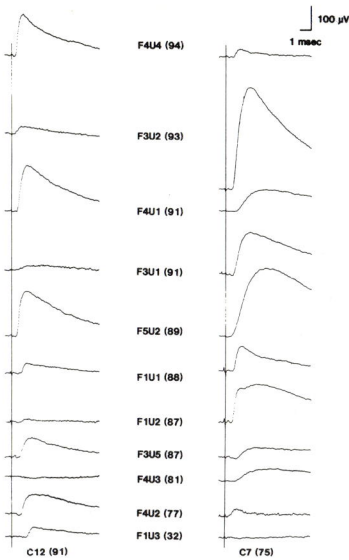

Fig. 6

Individual e.p.s.p.s evoked by
eleven afferent fibres in two
motoneurons in the same experiment. The conduction velocities
(m/sec.) of the motoneurons and
afferents are given in paren-
theses (after Clamann et al.
1985)

The connectivity matrix: In fig. 7 the entire set of connections studied in a single experiment - they number 165 - are shown in matrix
form. The afferents are arranged in columns, the motoneurons in rows.
Both are arranged in order of decreasing conduction velocities of
their axons. As already mentioned, axonal conduction velocity correla-
tes strongly with the size of the neuron. Spots represent active
connections, as defined by the presence of an individual e.p.s.p.
Empty squares mean, that no functional connections could be demonstrated. It is readily seen, that there are more active connections concentrated in the upper left quadrant while far fewer appear in the lower right. Thus a large afferent fibre tends to make more functional
connections than a small afferent fibre, and a large motoneuron seems
to have a higher probability of receiving functionally active connections than does a small motoneuron. Size related constraints, apparently place limits on the randomness of the connectivity matrix. It
is reasonable to infer, that the more extensive terminal arborizations
stemming from a large afferent fibre should have a greater likelyhood
of coming into close proximity with the dendrites of a motoneuron and,
thus, a higher probability of establishing connections with them.

Likewise, a large motoneuron with its dense dendritic meshwork
should have a greater chance of receiving connections than a small

motoneuron. Within these structural constraints, however, the connections are established at random.

| FOA 14 | F5 U3 100 | F1 U1 93 | F4 U1 91 | F1 U2 87 | F3 U1 87 | F3 U2 67 | F5 U1 66 | F5 U2 57 | F2 U1 53 | F3 U3 44 | F2 U2 37 |
|---|---|---|---|---|---|---|---|---|---|---|---|
| C 8 99 | ● | ● | ● | ● |  | ● | ● |  | ● | ● |  |
| C 6 97 | ● | ● | ● | ● | ● | ● | ● | ● |  |  | ● |
| C 11 97 | ● | ● | ● | ● | ● | ● |  |  | ● |  |  |
| C 4 97 | ● | ● | ● | ● | ● | ● |  |  | ● | ● |  |
| C 10 96 | ● | ● | ● | ● | ● | ● | ● | ● |  |  |  |
| C 13 93 | ● |  | ● |  | ● | ● | ● | ● |  |  |  |
| C 1 91 | ● | ● | ● | ● | ● |  |  |  | ● | ● |  |
| C 3 91 | ● | ● | ● | ● | ● |  |  |  | ● | ● | ● |
| C 12 90 | ● |  | ● |  | ● |  | ● | ● |  |  |  |
| C 5 89 | ● | ● | ● | ● | ● | ● | ● | ● |  | ● |  |
| C 2 88 | ● | ● | ● | ● | ● | ● | ● | ● | ● |  |  |
| C 15 88 |  |  |  | ● | ● | ● |  |  |  |  |  |
| C 14 84 |  | ● |  | ● | ● |  |  |  | ● |  |  |
| C 7 82 | ● | ● | ● | ● |  |  | ● | ● |  |  | ● |
| C 9 73 | ● | ● | ● | ● | ● |  |  |  |  |  |  |

Fig. 7

Connectivity matrix illustrating how the axonal conduction velocities of the motoneurons and afferent fibres in one experiment are related to the presence (●) or absence (blank squares) of functional connections between them (after Clamann et al. 1985).

## The concept of silent synapses

Failure to evoke an individual e.p.s.p. in a motoneuron does not necessarily indicate that there is no anatomical connection between the afferent fibre and the motoneuron. When HRP is injected into afferent fibres and homonymous motoneurons, evidence of anatomical connections can always be found. Some of these connections are apparently subject to what is called transmission failure. The absence of an individual e.p.s.p. indicates that the whole synaptic contact system, consisting of one to several boutons, fails to release transmitter; the synapses remain silent.

An analysis of the events at a single contact system indicates, that single synaptic boutons, or clusters of boutons, can remain silent for prolonged periods of time. Fig. 8 illustrates how the shape of an individual e.p.s.p. can change spontaneously. The e.p.s.p. with slow rise time changes to an e.p.s.p. with a short rise time and a composite time course of the falling phase. From the shape of the e.p.s.p. inferences can be drawn on the site of action of the synapses on the somato-dendritic tree (Rall, 1967). Synapses located at distal dendrites tend to produce slow rising and long lasting e.p.s.p.s while synapses locate close or at the soma produce short, fast rising e.p.s.p.s. With this knowledge the illustrated changes in the shape of the e.p.s.p. are readily explained by use of the schematic on the right side of fig. 8. It illustrates that within the same telodendron there exists a group of active dendritic synapses (filled triangles), which is responsible for the slowly rising potential and a group of somatic synapses (open triangles), which are at present silent. These silent connections can become active and contribute the steep rising phase and composite falling phase to the e.p.s.p. The mechanisms for their activation are not understood. Experimental evidence allows us to calculate that only about 25% of all synaptic boutons in these reflex arc release transmitter in response to an afferent impulse. The remaining 75% are normally silent. From afferent impulse to afferent impulse, a variable number of synapses can become active. This can be inferred from the amplitude fluctuations of individual e.p.s.p.s. It can be demonstrated, that the amplitude fluctuates in quantal steps, and evidence suggests that the quantas are best interpreted as the action of a single synaptic bouton (Jack et al. 1981).

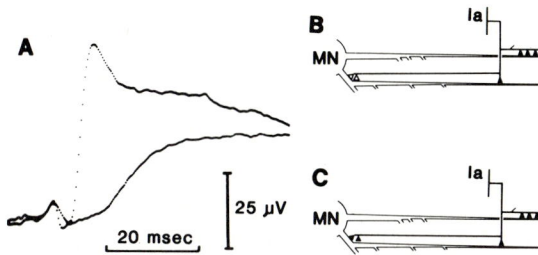

Fig. 8

A: Spontaneous change of a slow rising e.p.s.p. into a fast rising e.p.s.p. with a composite decay time course. This change could reflect the activation of previously silent synapses ($\triangle$) located at or close to the soma (B, C).

In summary it can be stated, that the complex structural network provides only a frame work, into which an ever changing functional network is embedded.


A synthesis


The course of analysis we have taken, began with the description of a deterministic behaviour of a fixed recruitment order of motor units and has led to the probabilistic laws governing the microscopic events at the level of the single synaptic boutons. What this fact demonstrates is a perfect example of a system as defined by Paul Weiss many years ago (Weiss, 1963). In short, he defined an ensemble of parts as a system, if the variance of the ensemble is significantly smaller than the sum of the variances of its parts. The essential feature of this definition is a criterion of stability which states that the behaviour of the system is on principle invariant to perturbations in its constituent elements. This definition implicitly assumes, that the behaviour of a system as a whole determines the behaviour of the constituent parts. Weiss coined the term macrodeterminism for this phenomenon, or as Popper named it "Verursachung nach unten" or downward causation (Popper, 1983; Campbell, 1974).

Now let me try to give a synthesis along the path of downward causation on the structural and functional phenomenons I have presented. I would like to do so by using a developmental concept for the formation of the random complex network. Suppose that during development and maturation of the spinal cord afferents grow into a group of motoneurons and form synapses at every possible chance encounter. The consequences for the formation of synaptic contact systems on a large and small motoneuron are illustrated in fig. 9. As we have seen, a large motoneuron produces a denser dendritic meshwork than a small motoneuron. The laws of chance predict that the larger motoneuron will receive a larger number of synaptic boutons than a small motoneuron from an afferent of a given size. This simple concept results in a rough proportionality between motoneuron size and the number of connections it receives from a particular afferent fibre.

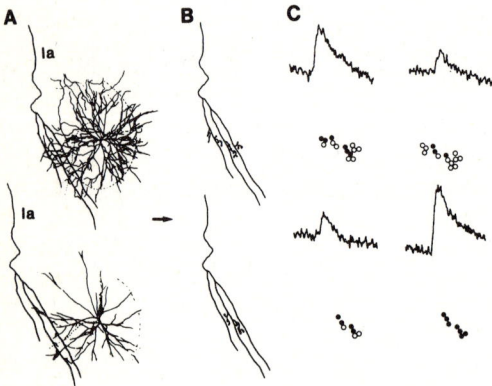

Fig. 9

Conceptual model on the random synapse formation and its consequences for the amplitude of the individual e.p.s.p. Further details in text (after Lüscher, 1985).

Of course, the same afferent fibre will make additional connections with other motoneurons in the neighbourhood. If we assume that a collateral branch of a certain diameter can supply a certain number of synapses, this population of synapses would be distributed to a group of motoneurons in proportion to their sizes. The path the afferent impulse has to travel along the axon in order to reach all the synaptic endings comprising a contact system on a single motoneuron, can readily be traced. This path is complex with many branch points in the case where a large afferent contacts a large motoneuron. It is a simple path for a contact system formed by small afferents on small motoneurons. Each such branch point may represent an uncertainty factor for the propagation of the action potential, because of an impedance mismatch between a parent axon and its daughter branches. The action potential therefore may reach only a fraction of all the synaptic boutons. The active subpopulation of boutons, symbolized with filled circles, consists of an ever changing number and grouping of the boutons, shifting from impulse to impulse. The number of synaptic endings activated by a particular impulse determines the amplitude of that particular individual e.p.s.p. The more profuse the axonal branching, the larger will be the number of silent synapses. In this way the extent of branching, i.e. the complexity of the telodendron determines the degree of functional connectivity between afferent and motoneuron. Thus the direct proportionality established during development between the number of synapses and the size of the motoneuron is functional redefined through the presence of silent synapses, and a rescaling occurs such that, on average, the largest motoneurons have the largest number of silent synapses. This results in a higher density of active synapses on small motoneurons, satisfying the biophysical requirements for the Size Principle of recruitment order while conforming to the anatomical observation that the density of synaptic endings is equal on small and large motoneurons.

This developmental concept may explain the emergence property of recruitment order in the complex neural network of the stretch reflex through downward causation. Why is it so difficult to follow the reverse path, trying to synthesize the system from the properties of the constituent elements ? Obviously, recruitment order is per definitionem a cooperative phenomenon. Indeed a single motoneuron can only fire an action potential but it can be recruited according to its rank order within a pool. If we know the rules of the game, we should be able to understand the systems behaviour from the properties of the constituent elements. May be, there are as many rules as there are elements of the system, because each element has its own individuality; not two are alike. From observation of the fluctuating individual e.p.s.p. it will never be possible to deduce the responsible neural structure, because infinite morphological variations together with transmission failure could produce the same functional behaviour. This inability to understand the function of the neural network through upward causation may not be a fundamental difference to physical systems. It may just demonstrate that we do not have the means yet to deal with the immense complexity of the central nervous system in the proper way.

Acknowledgments: The research summarized in this lecture was carried out in H.-R. Lüscher's laboratories at the Department of Physiology, University of Zürich. It was supported by a series of grants from the Swiss National Science Foundations, the Sandoz Stiftung, the Hartmann-Müller Stifung and the Roche Research Foundation. I am extremely grateful for their generous financial support of my research.

References:

Brown, A.G. and Fyffe, R.E.W.: The morphology of Ia afferent fibre collaterals in the spinal cord of the cat. J. Physiol. (London) 274: 111-127, 1978

Brown, A.G.: Organization in the Spinal Cord. Berlin, Heidelberg, New York: Springer-Verlag. 1981

Burke, R.E.: Group Ia synaptic input to fast and slow twitch motor units of cat triceps surae. J. Physiol. (London) 196: 605-630,1968

Burke, R.E., Walmsley, B. & Hodgson, J.A.: Structural-functional relations in monosynaptic action on spinal motoneurons. In: Integration in the nervous system. ed. Asanuma, H. and Wilson, V.J. pp. 27-45. Tokio: Igaku-Shoin, 1978

Campbell, D.T.: "Downward Causation" in Hierarchically Organized Biological Systems. In:Studies in the Philosophy of Biology, Ayala & Dobzhansky (eds.) 179-86, 1974

Henneman, E.; Somjen, G. & Carpenter, D.O.: Functional significance of cell size in spinal motoneurons. J. Neurophysiol. 28: 560-580, 1965a

Henneman, E. & Olson, C.B.: Relations between structure and function in the design of skeletal muscles. J. Neurophysiol. 28: 581-598, 1965b

Jack, J.J.B., Redman, S.J. & Wong, K.: The components of synaptic potentials evoked in cat spinal motoneurones by impulses in single group Ia afferents. J. Physiol. (London) 321: 65-96, 1981

Lüscher, H.-R., Mathis, J. & Schaffner, H.: A dual time-voltage window discriminator for multiunit nerve spike decomposition. J. Neurosci. Methods 7: 99-105, 1983

Lüscher, H.-R.: Selbstorganisation als Ordnungsprinzip im Zentral-
     nervensystem. Vierteljahresschrift Naturforschende Gesellschaft
     in Zürich, 128,: 167-180, 1983

Lüscher, H.-R.: Komplexität und Stabilität im Zentralnervensystem.
     Bulletin SAMW. 59-75, 1984/85

Popper, K.R.: Der Materialismus überwindet sich selbst. In: Das Ich
     und sein Gehirn, edd. K.R. Popper und J.C. Eccles, München 1983

Rall, W.: Distinguishing theoretical synaptic potentials computed
     for different soma-dendritic distributions of synaptic input.
     J. Neurophysiol. 30: 1138-1168, 1967

Weiss, P.A.: The living system: determinism stratified. In: Beyond
     Reductionism, ed. Koestler, A. & Smythies, J.R. pp. 3-55. London:
     Hutchinson, 1969

Westbury, D.R.: A comparison of the structures of $\alpha$- and $\gamma$-spinal
     motoneurones of the cat. J. Physiol. (London) 325: 79-91, 1982

Zucker, R.S.: Theoretical implications of the size principle of
     motoneuron recruitment. J. Theor. Biol. 38: 587-596, 1973

# PHYSICS OF THE BRAIN

Rodney M.J. Cotterill

Division of Molecular Biophysics
The Technical University of Denmark
Building 307, DK-2800 Lyngby, Denmark

## Abstract

The human brain consists of approximately one hundred thousand million cells, arranged in a variety of structures, the largest of which is the familiar neocortex. These cells, known as neurons, possess the vital property of excitability, which is dependent upon the differential diffusion characteristics of their bounding membranes. The cells receive and transmit electrochemical impulses through their numerous tentacle-like extensions, and the signals are passed from one cell to another by the chemical messengers called neurotransmitters, which diffuse across the narrow inter-cell gaps known as synapses. The efficiency of the transmission process is chemically modifiable, and this is believed to imbue the neural network with the ability to learn and remember.

The response to a variety of input patterns has been studied in a vector model assembly of interconnected neurons. The time evolution of the injected signal was followed, attention being paid to both its subsequent topology and phase. The model was realistic in that it included action potential impulses in the axon regions, statistically distributed synaptic delays, and electrotonic waves in the dendrites. Of particular interest were the frequency response of the system, and its dependence on the proportions of excitatory and inhibitory synapses. The relevance of the concept of coherence length was also critically examined, in such disparate contexts as association, autism and the primary visual processes in the retina. Coherence, and the more general issue of correlation, were also considered in connection with memory models, including those of the holographic type.

## 1. Introduction

There has recently been an enormous increase in activity regarding the structure and function of the brain. Much of this has been generated by the more general advances in biology, particularly at the molecular and microscopic levels, but it is probably fair to say that the stimulation has been due at least as much to recent advances in computer science. To accept this view does not mean that one is equating the brain to an electronic computer; far from it, most people involved in brain research have long since come to appreciate the considerable differences between the brain and the

computer. But the computer is nevertheless a useful device in brain science, because it permits one to simulate the functioning of a modest number of brain cells,
and the interactions between them. Several fascinating computer simulations of
brain function have indeed been reported in recent years, and it has been shown
that these are capable of such brain-like properties as association and pattern recognition [ 1-8 ].

Observed macroscopically, the brain gives little hint as to how it performs. In
the human, it has a familiar walnut-like appearance with its convoluted surface and a
clear division into two lobes. This major portion is the cerebral cortex, regions of
which have now been identified with particular faculties, as indicated in Figure 1.
Closer inspection reveals such distinct minor compartments as the cerebellum, located
at the lower rear; the hippocampus, pons, and thalamus, positioned centrally near the
junction with the spinal chord; and smaller structures like the hypothalamus, the pit-
uitary and pineal bodies, and the olfactory bulb. The functions of most of these small
regions are now reasonably well understood. The thalamus, for instance, appears to
act as a relay station.

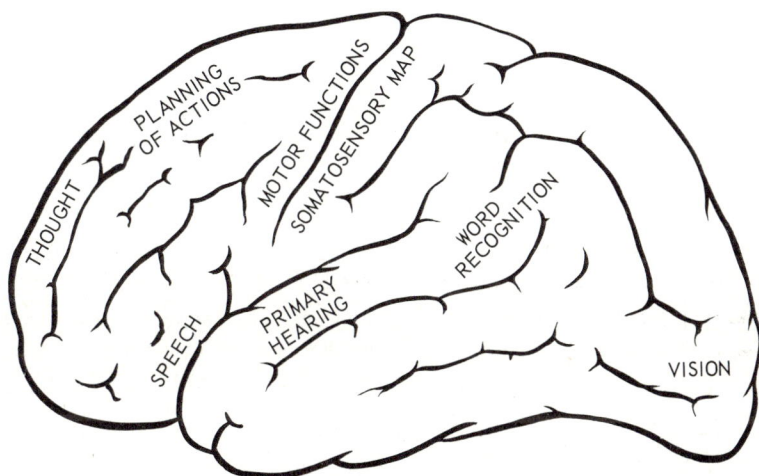

Fig. 1. Studies of patients who have sustained injuries to various
regions of their brains have permitted identification of cortical areas
linked to specific faculties. Some of the grosser divisions are indicated in this modified version of a picture due to Kohonen.

The brain consists of approximately one hundred thousand million nerve cells. A
typical cell has a shape not unlike that of a leguminous plant; a radish, say (see
Fig. 2). It has a reasonably well defined body, referred to as the soma, and a large
number of extended protuberances, which are referred to as processes. These latter,
extending outwards from the body like numerous tentacles, are of two types. There
are the dendrites, which are patterned rather like the limbs and branches of a tree,

and these have been found to carry signals towards the soma. These are known as the afferent processes. Then there is a single process extending from the soma over a distance that is often many times the diameter of the latter. This is the axon, which usually terminates in a similar branching pattern. It is referred to as an efferent process, and it transmits signals emanating at the soma, onwards towards other cells. Connections between the various nerve cells are highly common, and they are refer- red to as the synapses. They are small regions of near-contact, in which the signal from one cell to another is passed chemically, and is mediated by molecules known as neurotransmitters. Within a given nerve cell, the transmission of information from the dendrite towards the soma is electrochemical in nature and it is transmitted with an attenuation which is dependent upon distance and time. The transmission of a signal from the soma out along the axon is, on the other hand, of the all-or-nothing type known as an action potential, or nerve impulse. This signal propagates without appreciable attenuation, at a speed of approximately 20 metres per second, the dura- tion of the impulse lasting about one milli-second or so. Once the soma has given out such an impulse, it cannot be stimulated to generate a further impulse until a certain minimum time has elapsed, and this is referred to as the refractory period.

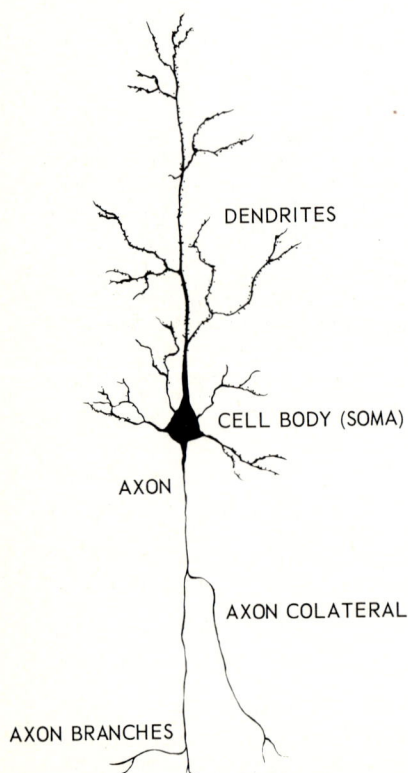

DENDRITES

CELL BODY (SOMA)

AXON

AXON COLATERAL

AXON BRANCHES

Fig. 2. The pyramidal cell, sketched here, is one of the prominent types of neuron found in the cortex. It takes its name from the shape of its soma, or cell body. Information, in the form of electrochemical waves, flows along the dendrites, towards the soma. If the voltage at the latter ex- ceeds a certain threshold value, an el- ectrochemical impulse is passed out along the axon. This signal, with its relatively high velocity (about 20 ms$^{-1}$) and all-or-nothing character, differs markedly from the slower, graded, and attenuated signals observed in the den- drites.

A given nerve cell can make upwards of a thousand synaptic contacts with neighbouring cells, so that the total number of synapses in the entire brain may be as high as $10^{14}$ or $10^{15}$. The time taken for the neurotransmitter molecules to diffuse across the synaptic gap is generally taken to be about one millisecond or so, and this is referred to as the synaptic delay. In practice there are several processes contributing to this mechanism, the first of these being the fusion of the small membrane-bounded sacks, which contain the neurotransmitter and are referred to as vesicles, with the pre-synaptic membrane. This causes the liberation of the neurotransmitter into the synaptic gap, and after the molecules have diffused across, they dock with receptor molecules which are able to generate the further electrochemical response in the dendrites referred to by the adjective electrotonic. These stages are shown in Figure 3.

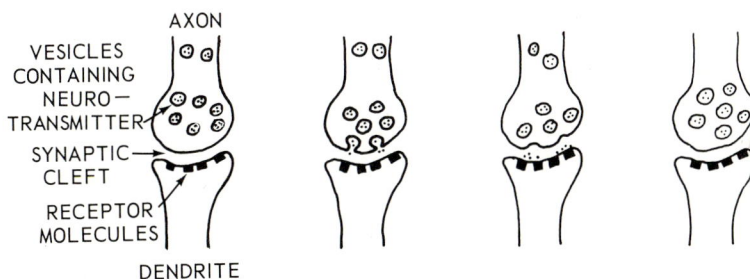

AXON

VESICLES
CONTAINING
NEURO—
TRANSMITTER

SYNAPTIC—
CLEFT

RECEPTOR
MOLECULES

DENDRITE

Fig. 3. The neurons in the brain are not in direct contact with one another, and the passage of information between these cells is a chemical process rather than an electrochemical one. The transfer occurs at structures known as synapses, which are primarily formed between an axon branch and the terminal region of a dendrite, (although axonal-somatic and dendro-dendritic synapses are also encountered). When the impulse arrives at the tip of the axon branch, it provokes fusion of vesicles, (which are membrane-bounded packets containing neurotransmitter molecules), with the pre-synaptic cleft, which is typically about 20 nanometres wide, and the neurotransmitter molecules drift across to the post-synaptic membrane and dock with receptor molecules. This initiates the graded electrochemical wave in the dendrite: the electrotonic response. The sequence of stages is here indicated from left to right.

Anatomical observations of the microstructure have revealed that many different types of nerve cells are present in the brain, and that they are linked up in a manner rather suggestive of electronic circuitry. The cells do thus not bear a relationship to the whole as do, for example, the atoms in a crystal. It is important to emphasize this in view of the recent emergence of spin-glass models of the brain [7,8], which take no account of the observed variety of cell types. The cerebral cortex, or neocortex as it is also called, consists of sheets of cells roughly 3 mm thick, and it is highly convoluted to permit its accommodation within the skull. There are indications of a subdivision within these sheets, groups of cells being lined up in columns of approximately 0.5 mm in diameter, lying perpendicular to the cortical surface. The density of synaptic contacts between the cells in a given column is rather high, whereas there are somewhat fewer synaptic junctions between the various columns, which are observed to make up a loose mosaic. It was this latter arrangement which suggested the structure for the model which will be described in a later section of this paper.

At this relatively early stage, one should not be impatient if reliable answers are slow in presenting themselves. Rather, it seems that this is the time for correctly formulating questions, and the following are offered as examples of questions which would seem to lie central to some of the most important issues. Why, for a start, are there roughly a thousand synapses per neuron as opposed to, say, ten or a hundred? Then again, why is it that one in some cases observes a reduction in the number of cells per topological layer, as, for example in the case of the visual system? The retina comprises roughly a billion cells, whereas there are only about a million cells in the optic nerve, which leads from the retina towards the lateral geniculate nucleus in the thalamus. A probably related question is: does the brain do calculus (i.e. integration or differentiation)? It seems likely that this is the case. Some of the recent spin-glass models require the existence of reciprocal synapses [7,8], for correct functioning, and this leads to the obvious question as to whether such an arrangement does occur in the cortex. Indeed this leads on to an even more fundamental question, namely whether the innermost recesses of the memory areas function in a vector or scalar fashion. And although the connection might not seem obvious, we could go on to ask whether we ever have any truly abstract thought? The point here is that if all our thoughts are related to one or another of the senses, the suggestion would be that the brain is always functioning in a vector manner. This, in turn, leads to what is a particularly fundamental issue, namely whether the interactions between the neurons occur in a coherent or incoherent fashion, and this particular issue re-emerges throughout the current paper; it will indeed be our main concern.

## 2. Primary visual processes at the retina

We turn now to the question of vision, and to the possibility that the classic observations of Hubel and Wiesel [9] can be taken as support for the idea that correct functioning in the brain is dependent on coherent excitation of various neurons. The observations in question were made on cells in the visual cortex of cats, using microelectrodes that were so fine that the activity of a single cell could be measured. The cats were anaesthetized with their eyes open, the controlling muscles having been temporarily paralysed so as to fix the stare in a specific direction. Hubel and Wiesel discovered that a given cortical cell can be specifically sensitive to a bar of light moving across a particular region of the cat's visual field, but only if the bar has a certain specific orientation and is moved in a certain direction.

These observations have been explained by assuming certain patterns of synaptic connections to the relevant cells in the lateral geniculate neucleus [9], the latter being a small knee-shaped region which is part of the thalamus, part of which acts as a sort of relay station in the visual pathway. Of particular interest here is the fact that Hubel and Weisel observed that there is a particular velocity of the moving light bar which gives the maximum response at the corresponding cell in the visual cortex. This most favourable velocity lies at around five degrees per second. From the geometry of the situation it is reasonably straight-forward to show that the speed of the image of the bar across the retina is equivalent to approximately one cell diameter during a time interval of about 10 milliseconds. This is a rather suggestive value, because it is comparable to typical electrotonic response times over typical dendritic lengths [ 10 ]. Indeed, these characteristic dimensions and times become even more interesting when we look at the underlying structure of the retina (see Figure 4).

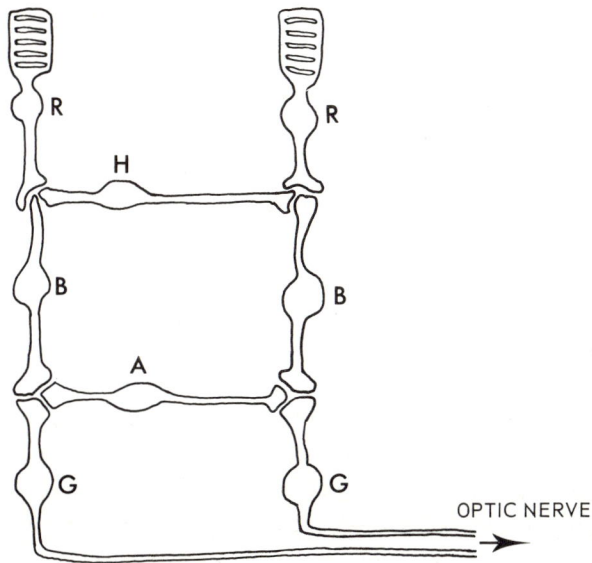

Fig. 4. Anatomical studies have established that the mamalian retina has an orderly structure composed of five different types of cell. In this schematic picture, these types are indicated by letters: R, for the receptors, which convert the energy of incident light photons into an electrochemical response; H, for the horizontal cells; B for the bipolar cells; A for the amacrine cells; and G for the gang-lion cells, the axons of which collectively form the optic nerve. The structure of the retina is somewhat surprising in that the incident light must pass the numerous cells of the other four types before it reaches the receptors (i.e. the light enters from below, in the figure).

There are five distinct cell types in this part of the eye: the receptor cells, which are responsible for converting the energy of the incoming photons into electrical activity; the horizontal cells; the bipolar cells; the amacrine cells; and finally the ganglions, which have a highly elongated shape, with their axons actually constituting the first part of the optic nerve. The receptor cells have their long axes lying normally to the surface of the retina, whereas the horizontal cells lie in the plane of the retina, and indeed form contacts between the receptor cells. These horizontal cells are rather special in that they have no well-defined directionality, and there is indeed no clear differentiation into dendritic and axonal extensions. It thus seems rather unlikely that these horizontal cells display action potential activity. Their responses are more likely to be of the electrotonic type, with the longer time constants associated with that type of function [ 10 ].

Let us suppose that the role of a particular horizontal cell is exclusively excitatory. We imagine that the moving bar of light falls first on one of the receptor cells, and then travels on in the direction of the next receptor cell down the line. Illumination of the first receptor cell elicits a response, which is passed along the plane of the retina by the horizontal cell. Because the electrotonic time constant of the latter is comparable to the above-stated 10 milliseconds, this electrotonic response will have precisely the timing required to produce reinforcement of the reaction of the second receptor cell, and so on. Because the response of the horizontal cell is certainly unilateral, this provides a mechanism which could underline the directionality observed by Hubel and Wiesel. It can, in fact, be looked upon as evidence supporting the idea of coherent excitation.

### 3.   A new computer model: the "pinch-out" effect

We will now describe a recently constructed computer model which aims at testing the idea of coherence, and at elucidating possible consequences of this mode of action. The model consists of a series of layers each consisting of the same number of cells, and with all possible combinations of the cells in two adjacent layers having unidirectional synaptic contacts. A single axonal input is assumed to feed into each synapse, and the latter is assumed to be followed by a single dendritic pathway to the subsequent somatic region. Because of the unidirectionality, an input pattern to the first layer, consisting of action potential pulses or a lack of these, will give rise to further patterns of firings and failures to fire, travelling down through the model layer by layer. Whether a particular synapse is excitatory or inhibitory is chosen by a random number generator, and this type of choice is also applied to the initial synaptic strengths, to their maximum values, and also to the time constants and maximum amplitudes of the electrotonic responses in the associated dendritic regions. Finally, the random number generator is also used to select a distributed set of values for the synaptic delays.

A study of the properties of this sytem, by computer simulation, has revealed several interesting modes of behaviour, one of which was certainly quite unexpected. The natural time-constant of such a system is determined by the minimum possible time lapse between successive pulses generated in a given cell, that is to say by the refractory period. A periodic input is given to the first layer of the system, and one then studies the successive generation of impulses in cells in the lower layers. If, for instance, some of the cells in the first layer are given impulses which are coincident with one another, it is found that there is a periodic generation of impulses in the lower layers, at the same frequency as the input frequency. But if cells in the first layer are given impulses which are temporarily offset from one another, a new phenomenon is observed, namely that after an initial transient period, it appears to be impossible for the cyclic state to maintain itself beyond a certain level in the system. This has been given the tentative name "pinch-out", and the phenomenon is illustrated in Figures 5a and 5b.

```
      1             8            14            20            21            61
 *000000*      .000000.      .000000.      .000000.      00000000      *000000*
 00000000      000000*0      000000.0      000000.0      000000.0      00000000
 00000000      00000000      0*000000      0.000000      0.000000      00000000
 00000000      00000000      00000000      00000000      00000000      00000000
 00000000      00000000      00000000      00000000      00000000      00000000
 00000000      00000000      00000000      00000000      00000000      0.0000.0
 00000000      00000000      00000000      00000000      00000000      .00000..
 00000000      00000000      00000000      00000000      00000000      000000..
 00000000      00000000      00000000      00000000      00000000      .0..00..
 00000000      00000000      00000000      00000000      00000000      *00..000
 00000000      00000000      00000000      00000000      00000000      000.0000
 00000000      00000000      00000000      00000000      00000000      00000000
 00000000      00000000      00000000      00000000      00000000      00000000
 00000000      00000000      00000000      00000000      00000000      00000000
 00000000      00000000      00000000      00000000      00000000      00000000

     82            200           421           436           781           961
 00000000      .000000.      *000000*      .000000.      *000000*      *000000*
 000000.0      000000.0      00000000      000000.0      00000000      00000000
 0.000000      0.000000      00000000      0.000000      00000000      00000000
 00000000      00000000      00000000      00000000      00000000      00000000
 00000000      00000000      00000000      00000000      00000000      00000000
 0.0000.0      0.0000.0      0.0000.0      0.0000.0      0.0000.0      0.0000.0
 .00000..      .00000..      .00000..      .00000..      .00000..      .00000..
 .0000...      .0000...      .0000...      .0000...      .0000...      .0000...
 .0...0*0      00..000.      00..000.      00..00..      00..000.      00..000.
 .00..000      000.0000      000*0000      000.0000      000*0000      000*0000
 00.*00.0      000.00.0      000.00.0      000.00.0      000.00.0      000.00.0
 nnnn0000      000000..      000000..      000000..      000000..      000000..
 00000000      000.*.00      000...00      000*..00      000...00      000...00
 00000000      00.0..0.      00.0..0.      00.0..0.      00.0..0.      00.0..0.
 00000000      .00.00.0      *00.00.0      .00.00.0      *00.00.0      *00.00.0
```

Fig. 5a. This computer model consists of fifteen topological layers, each comprising eight cells. Synaptic contacts are made between each cell in a given layer and all eight cells in the following layer. (There are thus 64 synapses between each pair of adjacent layers.) The transmission of information is unidirectional, from top to bottom, and the state of each cell is indicated by an 0, for the quiescent state, * for the moment of firing off an action potential, and a dot (.) is used if the cell is in its refractory period, which is of a standard length of 20 computational time steps. The electrotonic time constants were randomly selected, and uniformly distributed in the interval 1-75 time steps, while the synaptic delays were in this case all a standard single time step. The synapses were either excitatory or inhibitory, this being chosen at random. The periodic input pattern consisted of simultaneous firings of the first and eighth cells in the first layer, with a period of 60 time steps. In real time, one time step is about 0.5 ms. As can be seen from the situations at these twelve different instants, the network achieves a cyclic state, the period of which matches that of the input.

It is interesting to speculate on the possible advantage, to the brain, of such a phenomenon. It could indicate that signals are unable to penetrate to higher regions of the cortex, unless there is the requisite degree of synchronisation between the

```
     1          17          20          21          30          37
*0000000    .0000000    .0000000    00000000    0000000*    0000000.
00000000    000000*0    000000.0    000000.0    00000.0     0.0000*0
00000000    00000000    00000000    00000000    0.000000    0.000000
00000000    00000000    00000000    00000000    00000000    00000000
00000000    00000000    00000000    00000000    00000000    00000000
00000000    00000000    00000000    00000000    00000000    00000000
00000000    00000000    00000000    00000000    00000000    00000000
00000000    00000000    00000000    00000000    00000000    00000000
00000000    00000000    00000000    00000000    00000000    00000000
00000000    00000000    00000000    00000000    00000000    00000000
00000000    00000000    00000000    00000000    00000000    00000000
00000000    00000000    00000000    00000000    00000000    00000000
00000000    00000000    00000000    00000000    00000000    00000000
00000000    00000000    00000000    00000000    00000000    00000000
00000000    00000000    00000000    00000000    00000000    00000000
00000000    00000000    00000000    00000000    00000000    00000000


    61          200         421         450         781         990
*0000000    .0000000    *0000000    0000000*    *0000000    0000000*
00000000    000000.0    00000000    000000.0    00000000    000000.0
..00.000    .0000000    ..00.000    0.000000    ..00.000    0.000000
.0*0.000    .0.00000    .0*0.000    .0.00000    .0*0.000    .0.00000
000.0000    0.000000    0.0.0000    0.000000    0.0.0000    0.000000
0.0000..    000.0..0    000.0...    000.0..0    000.0...    000.0..0
.0000000    00000.00    00000.00    00000.00    00000.00    00000.00
000000.0    0000..00    0000..00    0000..00    0000..00    0000..00
.0000000    0000.00.    0000..00    0000..0.    0000..00    0000..0.
00000000    00..0000    0000*.0.    0000*.0.    0000*.0.    0000*.0.
00000000    000.00.0    .0.00000    .0.00000    .0.00000    .0.00000
00000000    000000.*    00.00000    00.00000    00.00000    00.00000
00000000    000...00    00000000    00000000    00000000    00000000
00000000    00*0*.0.    00000000    00000000    00000000    00000000
00000000    .00.00.0    00000000    00000000    00000000    00000000
```

Fig. 5b. An antiphase input to the model shown in Figure 5a gave a dramatically different response. The first cell in the first layer is made to fire at times 1, 61, 121 ....., while the eighth cell in that layer is made to fire at times 31, 91, 151 ..... As can be seen from these selected situations, information is able to reach the fifteenth layer only during an initial transient period. Thereafter, despite continuation of the input, nothing is able to penetrate beyond the eleventh layer. This phenomenon has been dubbed "the pinch-out effect".

various inputs to the first layer of cells. This, in turn, could mean that the appropriate regions act as a sort of coherence discriminator. Indeed, there is the suggestion that one could have a piece of cerebral hardware designed to respond to correlations between various inputs.

It is natural to speculate as to whether this type of circuitry could have relevance to what was stated earlier in connection with the primary visual processes at the retina. The general idea here would be that unless the signals were generated by the receptor cells at just the right time, no signal would ultimately reach the corresponding cells in the visual cortex. And because the correct synchronisation of these signals from the receptor cells would be dependent upon the light stimulus occurring at precisely the right time, this would give the velocity sensitivity accounted for in the previous section.

## 4. Fever in autistics [ 11 ]

We turn now briefly to the mental syndrome known as autism, which usually manifests itself during the patient's first four years of life. The autistic child is most often quite free from physical abnormality, and the chief symptom is a gross reticence or inability to interact with the environment. The patient appears apathetic to both people and objects, and in the early stages this can be mistaken for contentment. The condition apparently has an organic aetiology [12,13] with hereditary origins [14]. Possibly the strongest recent endorsement of the organic view comes from the widely observed, but inadequately documented, fever effect [15]. When autistics have a moderate fever, they invariably display dramatically more normal behavioural patterns, including a greater desire or ability to communicate. The effect appears to reach a maximum for fevers of around $2^{o}C$. It seems unlikely that such a modest rise could appreciably influence the rates of either the metabolic processes or the molecular diffusion involved in neural function. But temprature change of as little as $1^{o}C$ can markedly alter the fluidity of membranes [16], such as those which form the synapses and the neurotransmitter-charged presynaptic vesicles.

An increase in the fluidity of these membranes would lower the vesicle-synapse fusion time, and thereby decrease the synaptic delay. We have already seen how this latter quantity might control what could be called the neural coherence length, which is a measure of the degree of inter-neuron cooperativity. Lower synaptic delays would increase this length, and one could speculate whether the autistic fever effect indicates that there is a connection between coherence in the behavioural sense and actual physical coherence at the neural level. Equally intriguing is thepossibility that autism stems from a neural lipid composition profile which departs from the ideal.

It is clear that these issues would be amenable to investigation with the aid of computer models of the type described in section 3 of this paper, and such studies have recently produced some most interesting results. The model had thirty-two cells in each of its fifteen topological layers, in this case, and the parameter of interest was, of course, the synaptic delay. Figure 6 shows the dramatic result of changing the mean value of the latter from 0.5 ms to 2.0 ms. For the longer synaptic delay, the pinch-out effect is again observed.

This is a most intriguing result because it offers a particularly direct explanation of what might lie at the heart of the autistic syndrome. At normal body temperature, the patient's faulty lipid profile gives the synaptic delays that are too long and information is not able to traverse some critical part of the brain because of the pinch-out effect. During a sufficiently high fever, the increased synaptic membrane fluidity gives lower synaptic delays; the pinch-out effect disappears, the information gets through, and the patient appears to recover almost dramatically, only to go back into his or her invisible shell once the fever subsides.

```
00000000000000000.0.0.0.0.0.0.          00000000000000000.0.0.0.0.0.0.
0000.00000000000000000000000*00.0       000000000000000000000000000.0..0
00000000000000000000000000000.00        0000000000000000.000000000000000
00000000.0000000000000000000000         00000000.000000000000000.000000
000000000000000000000000000000000       0000000000000000000000000000000000
000000000000000000000000000000000       0000000000000000000000000000000000
000000000000000000000000000000000       0000000000000000000000000000000000
000000000000000000000000000000000       0000000000000000000000000000000000
000000000000000000000000000000000       0000000000000000000000000000000000
000000.,0000000000000.000000000         0000000000000000000000000000000000
.00.0000...00.00000.000..000.000        0000000000000000000000000000000000
00..0*0.0000.0.00.00..000.0...00        0000000000000000000000000000000000
000..0..000000...0000.....*..00.        0000000000000000000000000000000000
00000000..0000......0..0.*0..000        0000000000000000000000000000000000
0.0.0.00000.000.00.0.000000.0...        0000000000000000000000000000000000
```

Fig. 6. The pinch-out effect can also be induced by increasing the spread in the synaptic delays. The mean
delay in the model was a single time unit for the situation
shown at the left, in which information is clearly able to penetrate to the lower layers. But an increase of the mean synaptic delay to four time units gives pinch-out, as seen at the
right. The model comprised fifteen layers, each with 32 neurons,
and both pictures correspond to the situation at 800 time steps.

## 6. Memory

The final item to be covered in this brief review is something which has already
been touched upon, namely memory. It has been implicit in what has been advocated
that we endorse the idea, apparently first promulgated by Hebb [17], that memory
traces are stored via the agency of modifiable syanpses. Because of its dependence
on our central theme, coherence, the holographic theory of memory [18-22] is of
special interest, even though one recent article in this area [23] accords prime of
place to the glial cells rather than the syanpses.

As normally employed [24], the recording of a hologram occurs when the beam
scattered from an object interferes with a plane reference wave, to give a standing
wave pattern. Subsequent viewing of the holographic recording illuminated by the
reference wave alone reveals an image of the object. One of the great attractions of
the holographic process, in the context of memory, is that the entire object can be
imaged in this way, albeit at a lower resolution, if only a fraction of the hologram is
used. This is reminiscent of Lashley's observation of similar apparent total recall in
ablated animal brains [25].

It is actually implicit in the analysis of the holographic process that the reference
wave is not really required. There is an alternative mechanism, even though it is not
normally practicable. It arises in the following way. Let us imagine the object as
being composed of two parts A and B. Illumination of the composite, AB, with coherent light produces interference, and the resultant standing wave pattern can be recorded in the usual way. If that hologram is now illuminated only be light being scattered from A, an image of B will in principle be seen. But this would be very difficult
to achieve in practice because the wave emanating from A would not be plane, and the

slightest displacement of A would preculde the desired reconstruction. In the case of a system of neurons, however, the dendrites and axons function in a manner analogous to optical fibres, and there is no marked vulnerability to disturbance in this way.

Now a good argument could be made for the proposition that the primary function of the neocortex is to record correlations. Perhaps the modifications, be they in the synapses or the glial cells, are analogous to AB correlations in the above-described mode of holography. If that is the case, the system would have the highly desirable property, that stimulation of the appropriate region of the neocortex by input A would elicit a memory recall of B, and vice-versa.

This issue, too, has proved emminently amenable to study by the present computer model, and some insight has even been gained into another of the questions listed in the introduction, namely why there are so many synaptic contacts per neuron.

```
*0*0*0*0*0*0*0*0*0*0*0*0*0*0*0*0*0*0*0*0*0*0*0*0*0*0*0*0*0*0*0*0*0*0*0*0*0*0*0*0*0*0*0*0*0*0*0*0*0*0*0*0*0*0
00000000000000000000000000000000000000000000000000000000000000000000000000000000000000000000000000000000
00000000000000000000000000000000000000000000000000000000000000000000000000000000000000000000000000000000
00000000000000000000000000000000000000000000000000000000000000000000000000000000000000000000000000000000
00000000000000000000000000000000000000000000000000000000000000000000000000000000000000000000000000000000
00000000000000000000000000000000000000000000000000000000000000000000000000000000000000000000000000000000
00000000000000000000000000000000000000000000000000000000000000000000000000000000000000000000000000000000
00000000000000000000000000000000000000000000000000000000000000000000000000000000000000000000000000000000
00000000000000000000000000000000000000000000000000000000000000000000000000000000000000000000000000000000
```

```
00000000000000000000000000000000000000000000000000000000000000000000000000000000000000000000000000000000
00..0.000.0000..0.,0...0.00......0.0.00.......0000.0.0..00.00..000.00..000.0..0..0.000000000.0..000.000.0..000....00.00...0000.
0.0000.0000000.000000.0.00000000000000000000.0000*000000.0000000000.0000000000000.000.0000000.*00000.00000..00000.0*000000000
00000000000000000000000000000000000000000000000000000000000000000000000000000000000000000000000000000000
00000000000000000000000000000000000000000000000000000000000000000000000000000000000000000000000000000000
00000000000000000000000000000000000000000000000000000000000000000000000000000000000000000000000000000000
00000000000000000000000000000000000000000000000000000000000000000000000000000000000000000000000000000000
00000000000000000000000000000000000000000000000000000000000000000000000000000000000000000000000000000000
00000000000000000000000000000000000000000000000000000000000000000000000000000000000000000000000000000000
00000000000000000000000000000000000000000000000000000000000000000000000000000000000000000000000000000000
```

```
*0*0*0*0*0*0*0*0*0*0*0*0*0*0*0*0*0*0*0*0*0*0*0*0*0*0*0*000000000000000000000000000000000000000000000000000
00000000000000000000000000000000000000000000000000000000000000000000000000000000000000000000000000000000
00000000000000000000000000000000000000000000000000000000000000000000000000000000000000000000000000000000
00000000000000000000000000000000000000000000000000000000000000000000000000000000000000000000000000000000
00000000000000000000000000000000000000000000000000000000000000000000000000000000000000000000000000000000
0000000..000.00.000000000.000.00000000000.00000.0.000.0000000.00.0..000000.0.00.000000000.0..000000000.0.000000.00.00000000.000
00000000000000000000000000000000000000000000000000000000000000000000000000000000000000000000.000000
00000000000000000000000000000000000000000000000000000000000000000000000000000000000000000000000000000000
00000000000000000000000000000000000000000000000000000000000000000000000000000000000000000000000000000000
00000000000000000000000000000000000000000000000000000000000000000000000000000000000000000000000000000000
```

```
00000000000000000000000000000000000000000000000000000000000000000000000000000000000000000000000000000000
000.0.000..00.00.000.....000.0000...0.0000.0.0..00000.0.0..00000...0000.000000000.0.0000.000000.000....000000...0000.
0000000.00.000000000000000*0000..0000000.00....0000000000..00000000000C0000.0.0.0....0..0.000000.0000.00000000000..0.00.0..000.0
00000000000000000000000000000000000000000000.0000000000000000000000000000000000000000000000000000.000000000000
00000000000000000000000000000000000000000000000000000000000000000000000000000000000000000000000000000000
00000000000000000000000000000000000000000000000000000000000000000000000000000000000000000000000000000000
00000000000000000000000000000000000000000000000000000000000000000000000000000000000000000000000000000000
00000000000000000000000000000000000000000000000000000000000000000000000000000000000000000000000000000000
00000000000000000000000000000000000000000000000000000000000000000000000000000000000000000000000000000000
00000000000000000000000000000000000000000000000000000000000000000000000000000000000000000000000000000000
```

Fig. 7. Four different situations are depicted here, in the order in which they occurred. The model has been further stretched out, and now comprises ten layers, each with 128 neurons. The top picture shows the initial input (which has been designated AB in the text). The response in the second layer is shown in the second part of the figure, Hebbian learning having been in force during the intervening period. When only half the original input is now injected (designated A in the text), as seen in the third part of the figure, the second-layer response (bottom part of the figure) bears a striking resemblance to that earlier observed for the full AB input. The network is thus functioning like a hologram.

The model was further stretched out to include 128 neurons in each of the ten topological layers. The synchronously-applied pattern consisted of action potentials to every other cell in the first layer, the 64 other cells remaining quiet. There is of course no geometrical significance to such a pattern because all cells in a given layer are connected to all cells in the following layer, but the choice of this alternating sequence has the advantage of being easily recognizable. Now because there are, by definition, no prior connections before any of these cells are encountered, any convenient division of them can be looked upon as giving independent groups of cells, which could play the roles of the portions A and B discussed above. Let us, for example, consider the 64 cells lying to the left of the mid-point of the first topological layer as belonging to part A, and the 64 cells lying to the right as constituting part B.

The composite pattern AB, namely the above-described firing of every other cell, was injected into the first layer and the consequent firing pattern in the second layer was monitored. Hebbian learning [17] was introduced by strengthening those synapses lying between cells which showed correlated firings, and weakening those involved in anticorrelations [7]. After a suitable learning period, the model was then subjected to only a portion of the first input, namely what has been referred to as part A. As can be seen in Figure 7, the resultant firing pattern in the second layer was remarkably similar to that which was previously observed for an AB input. To be specific, the original AB input caused 60 of the 128 cells in the second layer to fire, while the post-learning A input provoked firing of 42 cells in the second layer. Moreover, with the exception of just one cell, all the post-learning firings in layer two occurred in cells which also fired before and during learning.

The learning and recall effects demonstrated by these computations are, of course, caused by the phase discrimination in the model, and it is not difficult to account for the disparity between the 60 pre-learning firings and the 42 post-learning firings. The firing of a second-layer cell requires the simultaneous arrival of electrotonic pulses from a number of first-layer cells. These latter can of course lie either in portion A or portion B of the first layer. Although the most likely situation will correspond to a fairly even distribution of contributions from A and B, there will be some cases where a disproportionately large number of cells in just one half of the first layer cause a second-layer firing. Although this remains to be investigated, it seems likely that such lop-sided distributions will become progressively less common as the number of synaptic contacts per cell increases.

7. Conclusion

The theme of this paper has been the possible role of coherence in brain function. It has been shown that several aspects of cortical processing might be predicated on such coherence, and that the resultant advantage to the brain could be considerable. But the experimental evidence for coherence is not yet particularly extensive. It is hoped that this article has sufficiently illuminated the issue that it will be perceived as a suitable object for study, by both experimental and computer techniques.

One of the central issues in neuroscience concerns the nature of higher process-ing in the cortex. The mechanisms underlying the early sensory processing are now becoming clear, whereas the manner in which the sensory information is subsequently handled remains a virtually total mystery. Because such target structures as the muscles and the glands have to be informed whether or not they are to respond to a given set of sensory inputs, the task of the brain's higher processing will, in the final analysis, be one of passing on or not passing on information, and this suggests an important role for the pinch-out effect. It seems appropriate, indeed, to close this article with the following question: *Is the pinch-out effect the mechanism whereby the brain discriminates between information to be transmitted to its target structures and information to be blocked or ignored ?*

## References

1. Kohonen, T., Acta Polytech. Scandinavica, EI 15, (1971); EI 29, (1971), Intern. J. Neuroscience, 5, 27-29 (1973).

2. Pellionisz, A. and Llinás, R., Neuroscience 2, 37-48 (1977).

3. Hogg, T. and Huberman, B.A., Proc.Nat.Acad.Sci. U.S.A., 81, 6871-6875 (1984).

4. an der Heiden, U. and Roth, G., Synergetics of the Brain (ed. E, Basar, H. Flohr, H. Haken and A.J. Mandell) Springer-Verlag, 1983.

5. Clark, J.W., Rafelski, J. and Winston, J.V., Physics Reports 123, 215-273 (1985).

6. Kohonen, T., Self-Organization and Associative Memory, 2nd Edition, Springer-Verlag, 1984.

7. Hopfield, J.J., Proc.Nat.Acad.Sci. U.S.A. 79, 2554-2560 (1982).

8. Kinzel, W., Condensed Matter 60, 205-213 (1985).

9. Hubel, D.H. and Wiesel, T.N., J. Physiol. (London) 166, 106-154 (1962).

10. Shepherd, G.M., The Synaptic Organization of the Brain, Oxford University Press, 1979.

11. The ideas expressed in this section are built on those expressed in: Fever in Autistics, R.M.J. Cotterill, Nature 313, 426 (1985).

12. Da Masio, A.R., Arch.Neurol. 35, 777-786 (1978).

13. Piggott, L.R., J.Autism Dev. Disorders 9, 199-218 (1979).

14. Folstein, S. and Rutter, M., Nature 265, 726-728 (1978).

15. Sullivan, R.C., J.Autism Dev. Disorders 10, 231-241 (1980).

16. Träuble, H., Tuebner, M., Woolley, P. and Eibl, H., Biophys.Chem. 4, 319-337 (1976).

17. Hebb, D.O., The Organization of Behavior, Wiley, 1949.

18. Julesz, B. and Pennington, K.S., J.Opt.Soc.Am. 55, 604-612 (1965).

19. Gabor, D., Nature 217a, 548-550 (1968); 217b, 1288-1291 (1968).

20. Pribram, K.H., Sci.Amer. 220, 1-14 (1969).

21. Greguss, P. Nature 219, 482-485 (1968).

22. Longuet-Higgins, H.C., Nature 217a, 104-105 (1968).

23. Nobili, R., Phys.Rev.A 32, 3618-3626 (1985).

24. Soroko, L., Holography and Coherent Optics, Plenum, 1980.

25. Lashley, K.S., Brain Mechanism and Intelligence, Dover, 1983.

# MODELS OF NEURAL NETWORKS

Wolfgang Kinzel
Institut für Theoretische Physik III
Justus-Liebig-Universität, D-6300 Giessen

## Abstract

A short introduction to simple models of neural networks is presented.
Information processing is based on attractors in configuration space.
Some recent results are outlined.

## 1. Introduction

Obviously our brain has very complex abilities of information processing which are related to memory, vision, speech, thinking etc.. It
seems to be obvious, too, that such properties cannot be described by
simple physical models or concepts. Clearly it is impossible to describe the brain by a few variables and their mathematical relations.

However, the functioning of the brain is based on real material, namely a huge network of nerve cells (neurons) and their connections (synapses). We have about $10^{10}$ neurons; on the average each of them is
directly connected to about $10^4$ others. On a small length scale the
neural net looks random and homogenous. The detailed structure of the
connections is presumably being developed by selection during the
learning process in the first few years of life /1/.

Since even simple tasks of the brain activate millions of neurons
which are widely distributed /2/ the information processing is a cooperative effect. Although much is known about the functioning of a
single cell and its synapses it is clear that from this knowledge
alone one cannot understand the properties of the brain. However, only very little is known about the cooperation of neurons. For instance
it seems to be completely unclear how information is encoded, stored,
retrieved and associated with previously learned patterns.

One approach to the understanding of the functioning of neural networks is the study of simple mathematical models which concentrate on a few essential mechanisms. For instance one may model a neuron by a number +1 or -1 which indicates whether the neuron is firing or is quiescent, respectively. The synapses may be modelled by numbers which indicate their strength and their type (excitatory vs. inhibitory). Hence one models a network by a set of units and connections which change their state in time by simple rules. Information processing is usually modelled by the motion of the activity patterns while learning is represented by a change of synaptic efficiences.

This approach leads to a parallel and distributed processing of information. Its main advantages are fault tolerance and content addressable memory. It is believed to be an important progress in the psychology of cognition /3/.

One may divide such connectionists models into two classes: Mappings and attractors. In fact mappings are known for more than two decades. In the simplest version they consist of two layers of input and output information which are connected by a learning matrix (Perceptron) /4/. However, only a limited class of input-output mappings can be realised by a single matrix of connections /5,6/. A straightforward improvement is the introduction of additional layers of units and connections between input and output /6,7/. Such structures have recently been studied in the context of text-to-speech converters /8/, selforganization of output representations /9/, development of feature detection /10/ and learning by mutation and selection /11/. Somewhat more complicated models of cellular automata type have been implemented in silicon hardware and are being used for speech recognition and motion detection /12/.

The second class of models uses attractors of the nonlinear cooperative dynamics of a large number of interacting units for information processing /13/. The computational properties are very similar to the ones of related mappings. The difference is that an external stimulus forces the network into an initial state which then moves into a final stable firing pattern. It is the existence of large number of stable states which may be related to biological properties.

Such attractors are easily understood if one notices the relations of networks to models of solid state physics and statistical mechanics /14,15,16/. Materials, too, consist of a huge number of units like atoms, molecules or magnetic moments which move cooperatively by their

mutual forces. The macroscopic behaviour of materials can be understood in terms of simple mathematical models which strongly simplify
the microscopic mechanisms.

Of particular interest are disordered magnetic materials with competing
magnetic forces which are called "spin glasses" /17/. For low temperatures such system have an infinite number of metastable spin configurations. Such states are attractors in phase space since they are local minima of the energy. Spin glass models have been discussed in the
context of distributed content addressable memory /16,18/.

In the following section networks with competing connections which
work as attractors in phase space are introduced. Their relation to
spin glasses is shown in Sec.3, and their computational properties are
demonstrated in Sec.4. Some recent results are mentioned in Sec.5.


## 2. The Mathematical Model

The model consists of a set of variables $S_i$ which can only take two
values +1 or -1. $S_i$ models the state of a spin (up or down), a neuron
(firing or quiescent) or a switch (on or off). The elements $S_i$ are connected by real numbers $J_{ij}$, which model the magnetic interaction, the
strength of the synapse or the resistance of a wire from site j to site i. The dynamics of the elements $S_i$ is described by the local field

$$h_i = \sum_j J_{ij} S_j \qquad (1)$$

which is the internal magnetic field at the spin $S_i$, the local electrical potential at the neuron or the current through the switch, respectively. The motion of the elements $S_i$ is defined by

$$S_i = \text{sign } h_i \qquad (2)$$

with sequential or serial updating of the elements $S_i$. For spin glasses Eq.(5) describes the relaxation of the magnetic energy at zero
temperature while neurons or switches change their state if the local
potential passes through a threshold value.

The dynamics, Eq.(2), is later used for pattern recognition of an associative memory. Learning or storing information needs an additional
dynamics on a much longer timescale. Namely for learning the values

$J_{ij}$ of the bonds are adapted to the presented information. But Eq.(2) is meant for fixed bond values $J_{ij}$.

## 3. Spin Glasses

Spin glasses are disordered magnetic materials like a gold-iron alloy. For low temperatures the magnetic moments of the atoms freeze into a random magnetic structure. Such materials are modelled by Eqs.(1) and (2). Since spin glasses have competing interactions, the bonds $J_{ij}$ are chosen to be randomly distributed /19/. Usually one takes a symmetric Gaussian with width $\Delta J$. A positive coupling $J_{ij}>0$ favors parallel alignment of the spins, i.e. $S_i S_j >0$, while $J_{ij}<0$ supports antiparallel orientation $S_i S_j <0$. In a dense network of bonds the spins cannot align to all of the bonds, this "frustration" effect leads to a complicated valley structure of the energy E in configuration space. E is given by

$$E = -\frac{1}{2} \sum_{ij} J_{ij} S_i S_j = -\frac{1}{2} \sum_{i} S_i h_i \tag{3}$$

Note that for magnetic models one has symmetric bonds $J_{ij}=J_{ji}$. Only then an energy Eq.(3) can be defined which drives the dynamics Eq.(2). This means that for each spin flip the energy E is decreased, hence the relaxation always leads to a local minimum of the energy as sketched in Fig.1.

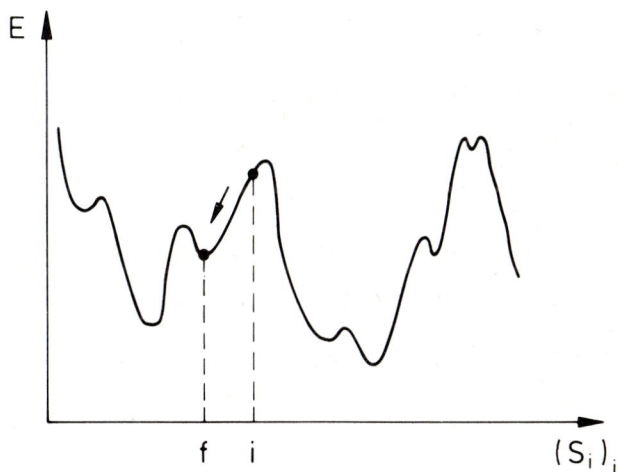

**Fig.1:**
Schematic view of the energy E as a function of spin configurations $(S_i)_i$ for spin glasses. An initial configuration i relaxes to the nearest local minimum which becomes the final stable state f.

A spin glass is a random symmetric network and from spin glass physics
one obtains two important properties:
(i)   The system always moves to a final stable state,
(ii)  the system has a huge number of stable states.

For a network with infinite ranged bonds the last point can even be
made quantitative /20/: The total number $N_s$ of stable states (for
which Eq.(2) is obeyed at each site) is given by

$$N_s \sim e^{\gamma N} \qquad\qquad (4)$$

where N is the number of spins and $\gamma \cong 0.2$. Hence already for small systems one has a large number of "valley bottoms"; for example for
N=100 spins one has:
(i)    $2^N \cong 10^{30}$ configurations $(S_i)_i$
(ii)   2        ground states (global minima) with energy
                 $E = -0.76...\Delta J \sqrt{N}$
(iii)  $\sim 10^8$  stable states (local minima) with energy
                 $E = -0.5...\Delta J \sqrt{N}$
(iv)   $\sim 10^4$  stable states with energy $E = -0.7...\Delta J \sqrt{N}$.

A random initial configuration relaxes into a state of class (iv) /22/.
Hence the dynamics Eq.(2) neither leads to the states of lowest energy
nor to the most probable states but to a state with an energy 8% above
the lowest valley. (Strictly this is true only for N→∞, for finite N
one observes a sharply peaked distribution on final energies /22/).
It is not known how many configurations relax to a given stable state
or whether all states of class (iv) can be reached by relaxation.


4. Associative Memory

Hopfield has pointed out how these properties of spin glasses are related to those of neural network models /16,18/. The problem is to define the bonds such that many given patterns are attractors of the dynamics, Eqs.(1) and (2). For random patterns this can be done by a
Hebb rule: If a pattern $(S_i)$, which is to be learned is presented to
the network, the bonds change locally by $\delta J_{ij} \propto S_i S_j$. For a set of M
patterns $\underline{S}^1 = (S_1^1, S_2^1, ..., S_N^1)$, $\underline{S}^2 = (S_1^2, S_2^2, ..., S_N^2)$, ....., $\underline{S}^M = (S_1^M, S_2^M, ..., S_N^M)$
which are learned with equal weight this gives

$$J_{ij} = \sum_{a=1}^{M} s_i^a s_j^a \qquad (5)$$

In fact Fig.2 shows that this procedure works: 30 patterns have been learned in a network of 400 neurons. Each pattern is a bottom of an energy valley with a large basin of attraction /21/. In a few updates per site a partial information relaxes to its learned counterpart.

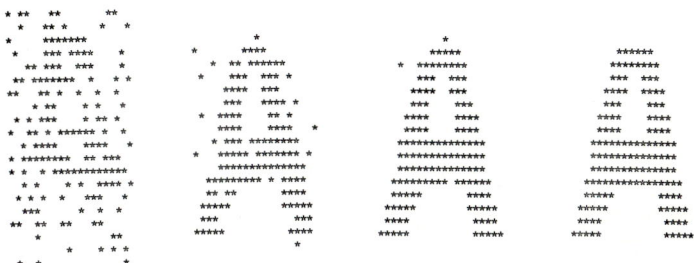

Fig.2:
In addition to 29 other (random) patterns the pattern "A" has been stored in the bonds of a network of 20x20 variables $S_i$. A star shows $S_i=+1$ while $S_i=-1$ is not shown. A pattern with 30% noise relaxes in 4 steps to its pure learned counterpart (from left to right).

Therefore, such a simple network has remarkable computational proper-ties. It works very differently from a current computer. Some main differences are listed in the following:

a) Storing and retrieving of information is a collective property of the network, no processing units and programs are needed.
b) Memory is distributed over the whole system, each bond contains a tiny piece of information about all of the learned patterns.
c) The information is retrieved by content, not by address.
d) The system is extremely fault tolerant, even if a large fraction of synapses and neurons are removed the system still works.
e) The information is retrieved very fast.
f) The dynamics of the neurons is parallel and asynchronous, each element may switch independently of the others.
g) The model also works if each neuron switches only with some proba-bility depending on the local field, such a stochastic dynamics even reduces retrieval errors.
h) All these properties are rather insensitive to model details.

These properties remind more of biological information processing than of present day computers. But of course one has to be aware that such a simple associative memory is still far away from complex information processing as in vision or speech recognition. Nevertheless there exists already an interesting application as a text to speech converter: a network which learns to read aloud /8/.

## 5. Some Recent Results

Finally we would like to mention some recent results with emphasis on work on which the author has collaborated.

## 5.1 Memory Capacity and Basin of Attraction

How many patterns can such a network store and how noisy patterns are recognized? The first question can be answered analytically using methods developed for the thermodynamics of spin glasses /23/: Up to M=0.144 N patterns are memorized with only little error, for $M/N \geq 0.144$ all stable states of the system are far away from the patterns $\underline{S}^a$, a=1,...,M. For $N \to \infty$ the system has a sharp transition at M/N=0.144.

The second part of the question concerns the size of the attraction basin. This part can be answered only numerically, at least at present /21/. The size of the attraction basin depends on the amount of stored patterns, i.e. on M/N. It increases with decreasing M/N.

The overloading of the network has recently been studied in the context of forgetting /24/.

## 5.2 Function vs. Connection

From the observed firing patterns of neurons one would like to derive the wiring architecture of the neural network /25/. Is this possible for the single models investigated above?

A measure for the correlations between the strength of the synapses and the state of the neurons at their ends is given by

$$\Delta = \frac{<J_{ij} \, S_i S_j>}{\sqrt{<(J_{ij} \, S_i S_j)^2>}} \tag{6}$$

where $<...>$ means an average over all bonds $J_{ij}$ and over all patterns $\underline{S}^a$. One finds for the Hopfield model /26/

$$\Delta = \frac{1}{\sqrt{M}} \tag{7}$$

Therefore, for a large number M of stored patterns $\Delta$ is very small and there is almost no correlation between synapses and stable states of the network. It is not possible to obtain the wiring diagram from the observed firing patterns.


## 5.3 Local Learning Rules

In the first few years of life a considerable fraction of synapses is dying off although a large amount of information is learned /1,27/. This motivated us to the following question: Can a random network learn by reducing bonds? In fact, starting from a spin glass and eliminating all bonds which are frustrated in one of the patterns the network memorizes a set of patterns /21,28/. This learning mechanism is local and works for correlated patterns, too.

However, the amount of patterns which are satisfactorily retrieved is rather small for this algorithm. On the other side, the Hebb rule Eq. (5) works for random patterns with the same amount of positive and negative bits, only. A similar but better learning scheme uses $S_i, S_j$ and the postsynaptic potential $h_i$ for the adaption of bonds $J_{ij}$ /29/. One possibility is given by

$$\delta J_{ij} \propto S_i S_j (c - S_i h_i) \tag{8}$$

with a given threshold c. One can prove that this algorithm converges if less than N linear independent patterns are learned by Eq. (8) in a sequential order /30/.

Such a local learning rule has several remarkable properties:

(i)  It works for any set of linear independent patterns.
(ii) After the learning is completed each pattern is imprinted with the

same strength, independently of how often Eq.(8) has been applied to it. In particular for each pattern and each neuron one
obtains the same strength $|h_i|=c$ of the neural potential.

(iii) Spurious stable states have a broader distribution of field
strengths $h_i$, hence the neurons get information on whether the
network is in a learned state or not.

These properties may be relevant for the functioning of real neural
networks.


## 5.4 Asymmetric Networks

The synapses derived from the Hebb rule Eq.(5) are symmetric. As was
pointed out in Sec.3, symmetry of the bonds guarantees final stable
states of the dynamics. On the other side, asymmetric random networks
behave very differently from symmetric ones (=spin glasses). The states remain wandering through configuration space in a chaotic way /31/
and the phase transition of spin glasses is destroyed by a small
amount of asymmetry /32/.

Synapses usually work in one direction, only. Therefore, one would
like to know whether the Hopfield model works for asymmetric bonds,
too. We have investigated this problem for the complete asymmetric
model /33/. The bonds are given by Eq.(5) except that one of the two
bonds $J_{ij}$ and $J_{ji}$ which is chosen randomly is set to zero.

Numerical simulations showed that many of the properties of the symmetric network hold for the asymmetric one, too. In particular we
find a discontinuous overloading transition at $M/N=0.075\pm0.005$.


## 6. Summary

A network of two-state threshold elements and competing bonds has interesting computational properties. For symmetric random bonds one
obtains a model for spin glasses. In this case the dynamics is driven
by an energy and always leads to final stable attractors. Spin glasses
have a huge number of stable states, but only few of them (still exponentially many) are attractors.

By a suitable adaption of the bonds such a network can store many

patterns simultaneously, i.e. each patterns becomes an attractor with
a large basin of attraction. Hence, the system becomes a content-ad-
dressable distributed memory. It has a parallel and asynchronous dynamics, it is extremely fault tolerant and works very fast.

The symmetric network undergoes a sharp change in its properties when
it is overloaded. This can be shown using analytical methods developed
for spin glasses. But also the asymmetric system which may be a more
realistic model for neural networks has computational properties very
similar to the symmetric one.

Several problems which are motivated from biology may be investigated
for such models. For instance, the network can learn arbitrary patterns by local adaption of synapses. It can forget by overloading or
saturation of synapses. A random network can learn by eliminating
connections. A layered system improves its computational properties
by choosing suitable output representations.

Acknowledgement

The author would like to thank S. Diederich, E. Domany, R.D. Henkel,
R. Meir, M. Opper and P. Spitzner for enjoying collaborations.

## References

/1/  See J.P. Changeux, this conference proceedings.
/2/  E.R. John, Y.Tang, A.B. Brill, R. Young and K. One, Science 233,
     1167 (1986).
/3/  S. Sutherland, Nature 323, 486 (1986).
/4/  K. Steinbuch, "Automat und Mensch" (Springer, Berlin 1963);
     F. Rosenblatt, "Principles of Neurodynamics" (Spartan Books,
     New York 1959);
     G. Palm, Biol. Cyb. 36, 19 (1980);
     T. Kohonen "Self Organization and Associative Memory" (Springer,
     Berlin 1984);
     H.D. Bloch, Rev. Mod. Phys. 34, 123 (1962).
/5/  M. Minsky and S. Papert, Perceptrons, (Cambridge. MIT Press 1969).
/6/  T.J. Sejnowski, P.K. Kienker and G.E. Hinton, Physica D, in press.
/7/  D.E. Rumelhart, G.E. Hinton and R.J. Williams, Nature 323, 533
     (1986).
/8/  T.J. Sejnowski and C.R. Rosenberg, John Hopkis Technical Report
     JHU/EECS-86/01.
/9/  E. Domany, R. Meir and W. Kinzel, Europhys. Letters 2, 175 (1986).
/10/ R. Linsker, Proc. Nat., Acad. Sci. USA (1986), in press.
/11/ M. Kerszberg and A. Bergman, proceedings of "Computer Simulation
     in Brain Science", Copenhagen 1986.
/12/ B.A. Huberman and T. Hogg, Phys. Rev. Lett. 52, 1024 (1984);
     K. Keirstaedt and B.A. Huberman, Phys. Rev. Lett. 56, 1094 (1986).

/13/ J.J. Hopfield in "Modelling and Analysis in Biomedicine", Nicolini (Ed.), World Scientific Press (Singapore, 1984).
/14/ W.A. Little, Math. Biosci. $\underline{19}$, 101 (1974).
/15/ P. Peretto, Biol. Cybern. $\underline{50}$, 51 (1984).
/16/ J.J. Hopfield, Proc. Nat. Acad. Sci. USA $\underline{79}$, 2554 (1982).
/17/ For recent reviews see: K.H. Fischer, Phys. Status Solidi $\underline{B\ 116}$, 357 (1983); $\underline{B\ 160}$, 13 (1985);
     K. Binder and A.P. Young, Rev. Mod. Phys. (1986).
/18/ W. Kinzel in: "Complex Systems", Ed. H. Haken (Springer-Verlag 1985), p.107.
/19/ S.F. Edwards and P.W. Anderson, J. Phys. $\underline{F\ 5}$, 965 (1975);
     D. Sherrington and S. Kirkpatrick, Phys. Rev. Lett. $\underline{35}$, 1792 (1975).
/20/ A. Bray and M.A. Moore, J. Phys. $\underline{C\ 13}$, L 469 (1980).
/21/ W. Kinzel, Z. Phys. $\underline{B\ 60}$, 205 (1985); $\underline{B\ 62}$, 267 (1986).
/22/ W. Kinzel, Phys. Rev. $\underline{B\ 33}$, 5086 (1986).
/23/ D.J. Amit, H. Gutfreund and H. Sompolinsky, Phys. Rev. Lett. $\underline{55}$, 1530 (1985);
     A. Crisanti, D.J. Amit and H. Gutfreund, Europhys. Lett. $\underline{2}$, 337 (1986).
/24/ J.P. Nadal, G. Toulouse, J.P. Changeux and S. Dehaene, Europhys. Lett. $\underline{1}$, 535 (1986); Forgetting by saturation of synapses is discussed by G. Parisi, J. Phys. $\underline{A\ 19}$, L 617 (1986).
/25/ J. Krüger in "Complex Systems", H. Haken (Ed.), (Springer Verlag 1985), p. 71.
/26/ W. Kinzel, unpublished.
/27/ J.P. Changeux and A. Danchin, Nature $\underline{264}$, 705 (1976).
/28/ J.L. van Hemmen and A.C.D. van Enter, Phys. Rev. $\underline{A\ 34}$, 2509 (1986).
/29/ S. Diederich, M. Opper, R.D. Henkel and W. Kinzel, to be published in the proceedings of the conference on "Computer Simulation in Brain Science", Copenhagen, Aug. 1986.
/30/ S. Diederich and M. Opper, to be published.
/31/ G. Parisi, J. Phys. $\underline{A\ 19}$, L 675 (1986).
/32/ J. Hertz, S. Solla and G. Grinstein, to be published.
/33/ P. Spitzner and W. Kinzel, unpublished.

# NONABELIAN NEURODYNAMICS

K.Hepp and V.Henn

Physics Dept.,ETH,and Neurology Dept.,University,Zuerich,Switzerland

Despite of its importance the human central nervous system (CNS) has remained impossible to understand in quantitative terms. There are many parallel pathways of contemporary research: Computational neuroscience tries to find efficient algorithms by which the brain could solve complex problems. One other possibility is to study simpler systems, like the brain of the sea-snail Aplysia [1], or simple subsystems of the CNS of higher mammals, in which there are less degrees of freedom, easier access for experimentation, and a clear understanding of the tasks. We are following the latter pathway and study the interaction between vision, motion detection and eye movements and try to relate our findings in the Rhesus monkey with human patho- logy. Fig.1 shows some of the relevant areas and connections of this

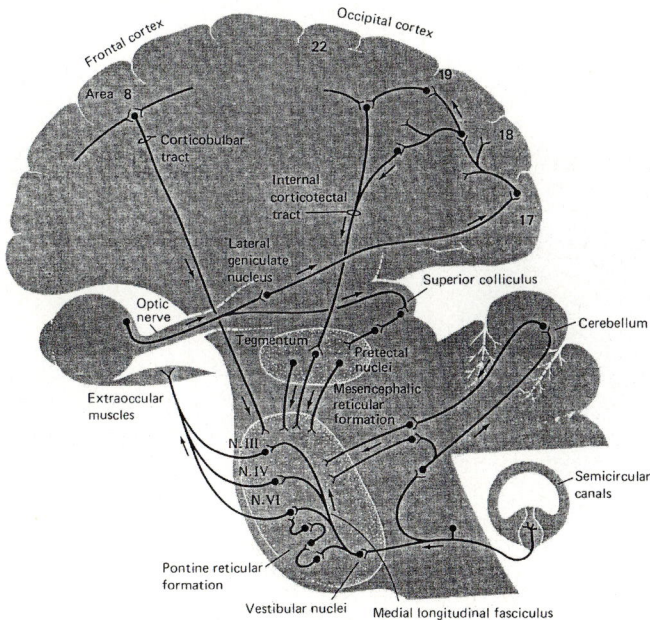

Fig.1  Visuo-vestibulo-oculomotor pathways (from [2])

subsystem. One sees that the "visuo-vestibulo-oculomotor interactions" cross many times the cortex, the cerebellum and the brainstem without being too heavily involved with information processing in the spinal cord,from which it is hard to record neural activity in alert mammals.

Our subsystem has still very many degrees of freedom: 6 for the movement of the head as a rigid body, which is sensed by the vestibular organ, 3 for the rotations of each eye, taken as spheres with centers fixed in the head, and many thousands for the visual patterns in the receptor arrays of the retinae. However, for important tasks the CNS matches the coordinate systems, so that the transformations become simpler. Fig.2 shows this for the compensatory eye movements which maintain the image of the world on the retina stationary even during completely unpredictable head movements.

The rotation of the head is sensed by the semicircular canals, which form two mirror symmetric pairwise almost orthogonal circular accelerometers, and by visual flow patterns, which activate in

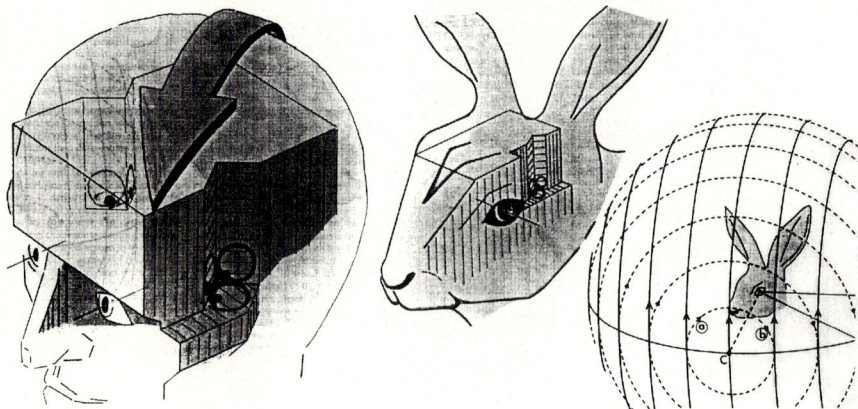

Fig.2  The orientation of the semicircular canals in relation to head rotation and visual flow patterns (from [3])

the subcortical visual system populations of neurons according to the same axial orientation as the canals (see e.g. [4]). Both patterns are combined in the vestibular nuclei (see e.g [5]) of the brainstem in a complementary way, the accelerometer signal from the canals for the higher and the visual flow input for the lower frequencies of head movements. In the "vestibulo-optokinetic ocular reflex" (VOKR) these signals induce compensatory rotations of the eyes in the opposite direction. Fig.3 shows the arrangement of the human eye muscles. As one sees from Fig.4 the rotation axes of 3 eye muscle pairs of each

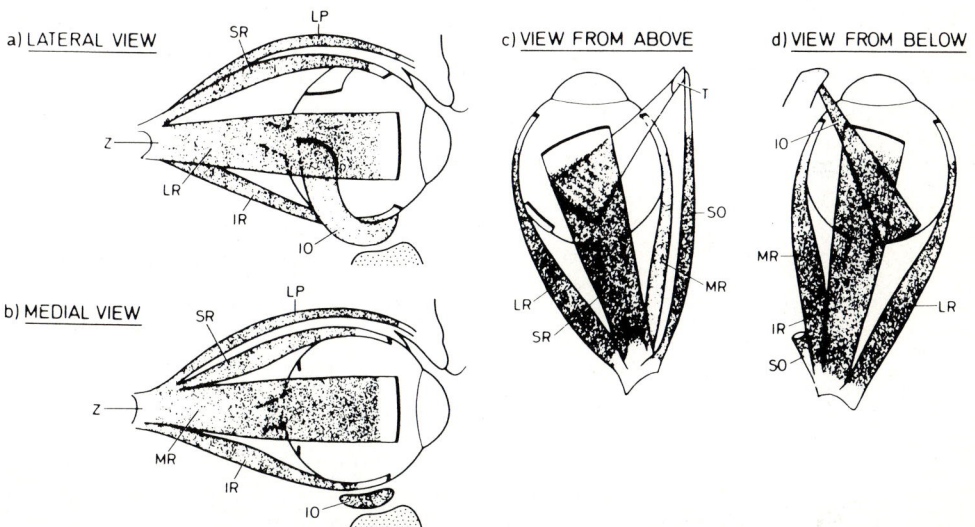

Fig.3  Human eye muscles (from [6])

eye are again pairwise approximately orthogonal and parallel to corresponding canal axes.

Another example of matching coordinate systems is the perifoveal region on the retina relative to the periprimary range of eye positions. The former is almost an Euclidean vector space of directions of point targets relative to the fovea. In a neighborhood of the primary position of the eye (see Fig.5) conjugate (i.e. parallel) fixations and saccades (i.e. rapid refixations) of the eye can be described in a 2-dimensional subspace of the tangent vector space of the rotation group SO(3), since by Listing's law the torsion is uniquely determined by the direction of the eyes (see below). Saccadic eye movements, by which the fovea "grasps" the target point along a stereotyped almost straight trajectory, can be described as 2-vectors, since they depend in the periprimary range only on the differences between initial and final eye positions which satisfy Listing's law. In the superior colliculus (see Fig.1) there is a visual map in which neurons are activated in relation to localized visual stimuli on the retina. In parallel to this map there is a saccadic target map in which neurons are activated, if a point in the visual map has been selected as target for a saccade. For the spatio-temporal transformations in this task see e.g. [8].

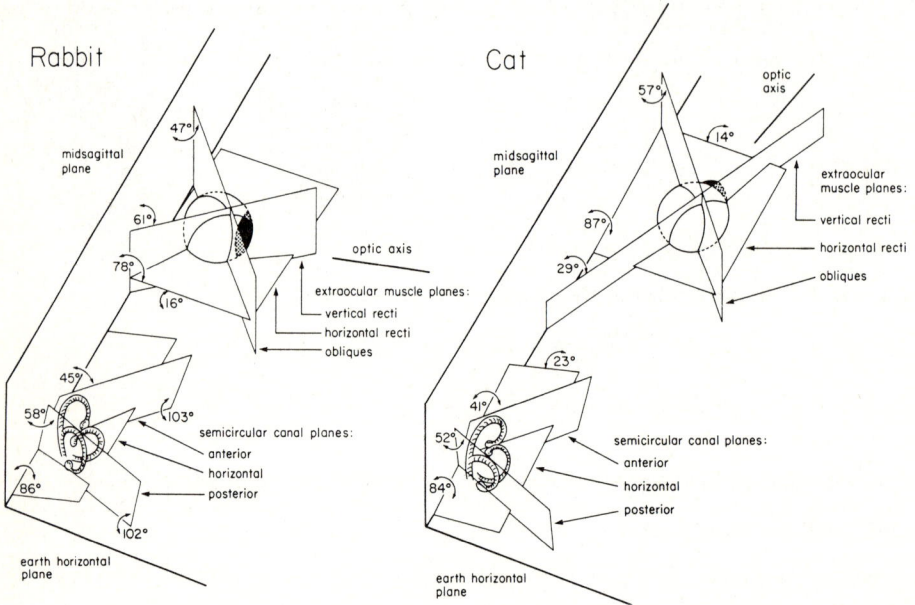

Fig.4 Orientation of semicircular canals and eye muscles (from [7])

It is, however, wrong to assume that sensorimotor transformations in the CNS are effected by separate channels, each being a sequence of maps and switched by context. Our behavioral repertoire is too large and flexible for this "grandmother to grandmotor" picture, which is

167

too simplified even for Aplysia. One finds a graded superposition of
channels, which the neurophysiologist calls "distributive" and for
which neurons with many thousands inputs and outputs are very efficient.
Even for visually evoked saccades, for instance, one can by clever
experiments completely dissociate the activity in the visual and saccadic maps of the superior colliculus. Furthermore, rapid saccade-like
eye movements can also occur in relation to head movements in the re-
setting of the eye during compensatory movements, when the eye approaches the border of the oculomotor range, and during orienting eye-head
movements, where the eye is brought foreward during the turning of the
head. For such tasks one needs a 3-dimensional control of saccades in
relation to the head coordinate system. We are trying to work out the
interaction of the different channels in the visuo-vestibulo-oculomotor
system, and we are delighted that a geometrical and mechanical understanding of the transformations, as pioneered by Mach [9] and Helmholtz
[10] and more recently by Robinson (see e.g.[11]), is extremely helpful
for choosing the right parameters in an otherwise completely incom-
prehensible experimental situation. As an example we shall analyse
mathematically the organization of the VOKR in 3 dimensions.

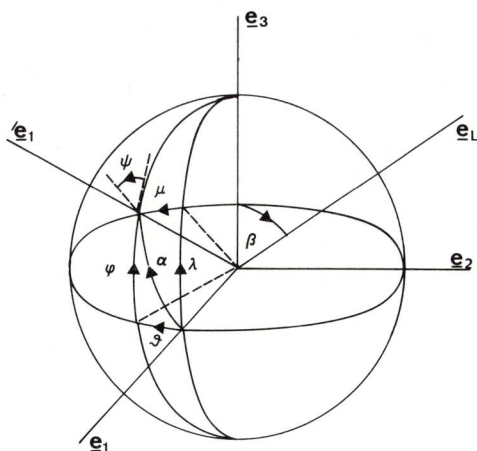

Fig.5  Fick and Listing coordinates

Let us choose a right-handed head-fixed "cranial" coordinate system with its origin in the center of rotation of the eye and oriented with the 1-axis $\underline{e}_1$ in frontal, 2-axis $\underline{e}_2$ in lateral (through the centers of rotation of both eyes), and 3-axis $\underline{e}_3$ in upward direction (see Fig.5). An eye-fixed "retinal" coordinate system with basis $'\underline{e}_1$, $'\underline{e}_2$, $'\underline{e}_3$ is chosen to coincide with the cranial system in primary position (to be defined below). Every orientation of the retinal relative to the cranial system within the oculomotor range is a possible eye position which is uniquely given by the $R \in SO(3)$ such that $'\underline{e}_n = R\underline{e}_n$. Fick coordinates have been used most frequently [11],[12] for eye rotations. If one chooses positive horizontal and vertical eye direction angles $\vartheta$, $\varphi$ to the right and up and torsion $\psi$ according to Fig.5, one first rotates around the $\underline{e}_3$-axis with $-\vartheta$ (by the right hand rule), then around the new $\tilde{\underline{e}}_2$-axis with $-\varphi$ in order to obtain the $'\underline{e}_1$-axis as eye direction and any torsion by a further rotation with $\psi$ around the $'\underline{e}_1$-axis. If the head is upright and stationary, Listing's law states that the torsion $\psi$ during fixation is determined by eye direction :

$$\sin\psi = \frac{\sin\vartheta \ \sin\varphi}{1 + \cos\vartheta \ \cos\varphi} \tag{1}$$

In humans (1) is rather well satisfied [12]. Listing's law is equivalent to requiring that the rotation $R = R(\underline{e}_L, -\alpha)$ of any fixation position has the axis $\underline{e}_L = \sin\beta \ \underline{e}_2 + \cos\beta \ \underline{e}_3$ in the 2-3-plane. Then $\beta$ and the rotation angle $\alpha$ satisfy $\sin\alpha \ \cos\beta = \sin\varphi$ and $\cos\alpha = \cos\vartheta \ \cos\varphi$. Primary position can experimentally be defined as the direction from which purely horizontal and vertical deviations introduce no static torsion.

Eye positions need not to be infinitely precise for accurate vision. During visually evoked saccades to a target at 10 deg monkeys make 1 deg errors (i.e. 10%). Using $R(\underline{e}, \beta) = \exp S(\beta\underline{e})$, where $S(\underline{x})\underline{y} = \underline{x} \wedge \underline{y}$, one can estimate the difference between $R(\underline{e}, \beta)$ and its "tangent vector"

$\beta \underline{e}$, using the Euclidean norm $\|R\| = sqr(\sum_{m\mu} R^2_{m\mu})$ , by

$$\| R(\underline{e},\beta) - 1 - S(\beta\underline{e})\|/\|R(\underline{e},\beta)\| \leq \|S(\beta\underline{e})\|^2/\sqrt{12} = |\beta|^2/\sqrt{3} \qquad (2)$$

Then a physiological definition of the "periprimary range" could be the range of $\beta$ , where $R(\underline{e},\beta)$ can be approximated by $1 + S(\beta\underline{e})$ in the sense of (2) up to 10%, and where for any 2 positions $R_1 = R(\underline{e}_1,\beta_1)$ and $R_2 = R(\underline{e}_2,\beta_2)$ the rotation $R_2 R_1^{-1}$ from $R_1$ to $R_2$ can also be approximated by $1 + S(\beta_2\underline{e}_2 - \beta_1\underline{e}_1)$ up to 10%. For $\beta_1 = \beta_2$ one obtains by this definition a periprimary range of more than 14 deg radius where saccades obey the law of vector addition up to at most 10% error (see Fig.10). Typically in an outdoor environment 86% of the saccades of three human subjects were 15 deg or less in magnitude [13].

There has been some historical dispute which coordinates are most natural for describing eye movements from within the nervous system. Angular velocity $\underline{w}$ of the eye in the cranial system is related to the changes $\dot{\vartheta} = d\vartheta/dt$, $\dot{\varphi}$ and $\dot{\psi}$ of the Fick angles by $\underline{w} = \dot{\vartheta}\,\underline{e}_3 + \dot{\varphi}\,\tilde{\underline{e}}_2 + \dot{\psi}'\underline{e}_1$ or by

$$\begin{pmatrix} \dot{\psi} \\ \dot{\varphi} \\ \dot{\vartheta} \end{pmatrix} = \begin{pmatrix} \cos\vartheta/\cos\varphi , & -\sin\vartheta/\cos\varphi , & 0 \\ -\sin\vartheta , & -\cos\vartheta , & 0 \\ \cos\vartheta\tan\varphi , & -\sin\vartheta\tan\varphi , & -1 \end{pmatrix} \begin{pmatrix} w^1 \\ w^2 \\ w^3 \end{pmatrix} = T(\vartheta,\varphi)\,\underline{w} \qquad (3)$$

For 1-dimensional movements, where $w^1 = w^2 = 0$ , $\dot{\vartheta} = -w^3$ is solved by quadrature $\vartheta(t) = \vartheta(s) - \int_s^t w^3(r)dr$, or in engineering terms by an integrator without position feedback. The integration of (3) requires position feedback. This complicates enormously the design and analysis of a neural network to implement the VOKR [14]. Here a head angular velocity signal $\underline{w}_H$ is generated in the vestibular nuclei from the canal and visual input. A fully compensatory eye rotation would require in the cranial system $\underline{w}(t) = \underline{w}_H(t)$ and therefore the integration of (3) to derive an eye position signal. If one replaces $T(\vartheta,\varphi)$ by $T(0,0)$, then $\|T(\vartheta,\varphi)-T(0,0)\|/\|T(0,0)\| \leq 10\%$ in compensatory movements (allowing for the sloppiness of the torsional system [12]) in about one half of the periprimary range (Fig.10), but it is unclear what functional use the CNS should make with a neuronal signal proportional to the Fick angles. We shall see that by a clever geometrical arrangement the lengths of the eye muscles are much better coordinates for the eye position, which the CNS can use to solve the integrator problem.

Look at Fig.3 for the anatomy of the extraocular muscles. The four recti (LR,MR,SR,IR) and the superior oblique (SO) originate at the back of the orbita, while the inferior oblique (IO) has its origin at the nasal rim of the orbit. Note the broad insertions of all muscles at the globe which, together with ligaments and fat tissue, do not allow them to take the shortest path over the globe. Simonsz [15] has studied the paths of the recti using computed tomographic scans in a plane perpendicular to the muscle cone and found in humans no consistent sideways displacement of the horizontal recti for 30 deg up and down and of the vertical recti for 30 deg right and left from primary position. A geometrical analysis of possible muscle paths over the globe has been made by Robinson [11], who concluded that in the cranial coordinate system the directions of all eye muscle moment vectors change little for all eye positions in the oculomotor range. Using recent data by Simpson et al. [16] on the position of the insertions

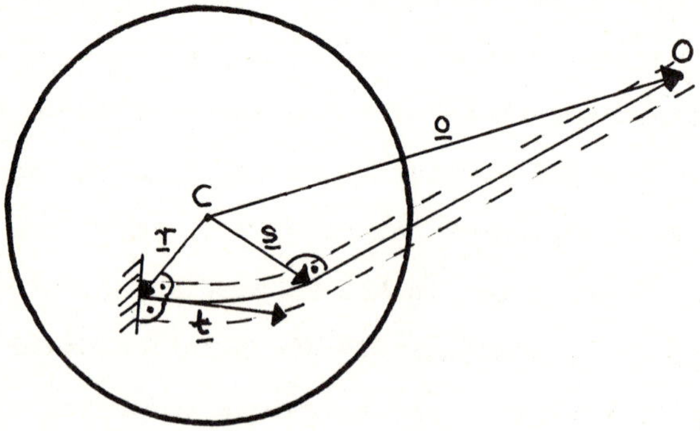

Fig.6  Hypothetical eye muscle path

Fig.7 (next page)  Iso-length curves for the eye muscles of the Rhesus monkey in Listing positions

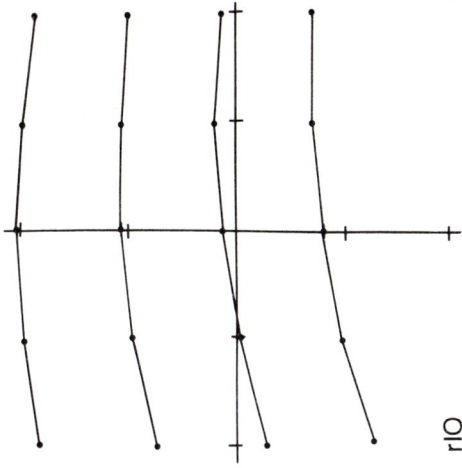

rIO

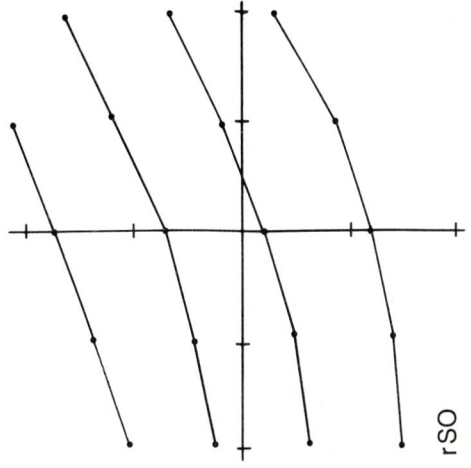

rSO

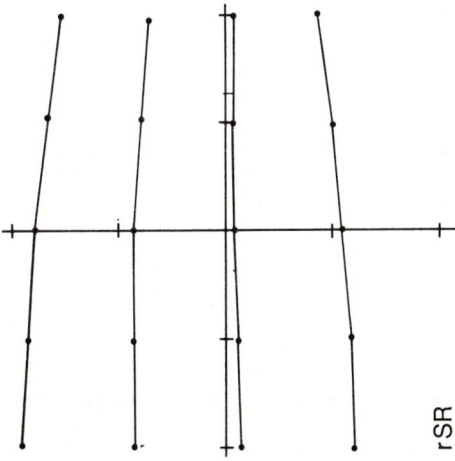

rSR

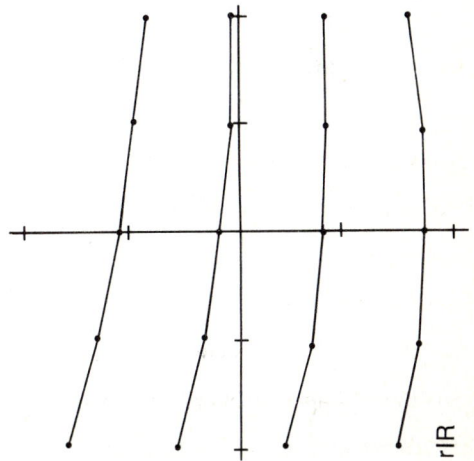

rIR

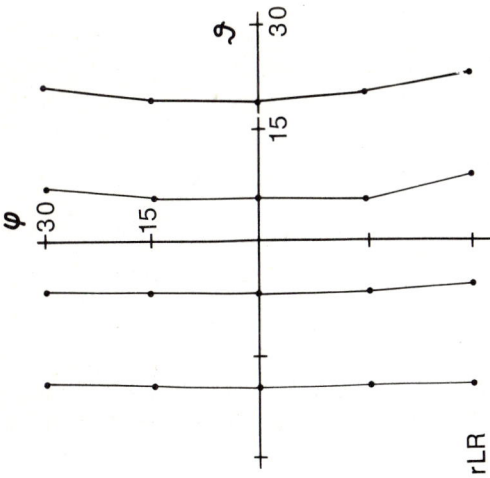

rLR

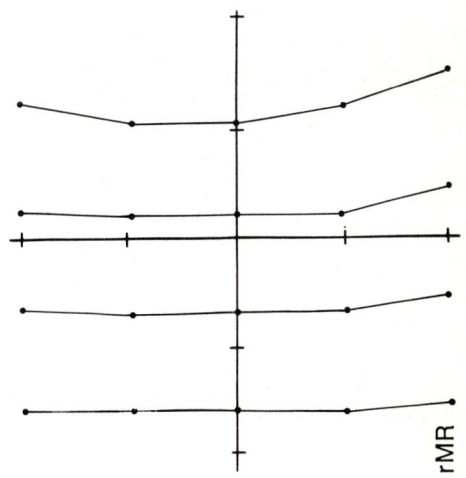

rMR

in primary position $\underline{r}_o$ and of the origins $\underline{o}$ of all muscles in the Rhesus monkey, we have computed their curves of equal length ("iso-length curves") for eye positions $\vartheta$, $\varphi$ satisfying Listing's law. Fig.7 was obtained under the assumption that the muscles follow the arc of a non-great circle which starts in the direction $\underline{t}=R\underline{t}_o$ of the tendon and stays in the plane through $\underline{t}$ and $\underline{o} - \underline{r}$, $\underline{r}=R\underline{r}_o$. According to J.I.Simpson (unpublished) $\underline{t}_o$ of the lateral rectus was rotated by 10 deg anticlockwise relative to a great circle path (see Fig.6).

These iso-length curves show a remarkable qualitative similarity to the iso-frequency curves of motoneurons in the same muscles of the Rhesus monkey [17]. During fixations and slow eye movements the firing rate $f$ of every motoneuron of the horizontal recti satisfy in good approximation the Robinson equation (see [18] and Fig.8)

$$f(t-\tau) = u + v\,\vartheta(t) + w\,\dot{\vartheta}(t) \tag{4}$$

with a time delay $\tau = 5$ ms and constants u, v, w characteristic for each motoneuron. Iso-frequency curves are obtained by plotting during fixation the positions in the $\vartheta-\varphi$-plane, where a recorded motoneuron fires at a certain rate. In Fig.9 these curves are shown for MR, IR, SO and for upwards pulling muscles (SR,IO). According to (4) the iso-frequency curves of MR motoneurons should be parallel to the vertical axis, which is true in the adducting half plane. Vertical motoneurons (see Fig.9) do not satisfy (4) with $\vartheta$ replaced by $\varphi$. We conjecture, from the similarities between the iso-length and iso-frequency curves for fixation, that

$$f(t-\tau) = u + v\,l(t) + w\,\dot{l}(t) \tag{5}$$

is a better approximation for the firing rate of all motoneurons along arbitrary slow movement trajectories, even far away from Listing positions, if the angular coordinate in (4) is replaced by the length $l(t)$ of the corresponding muscle. We are presently checking (5) using a 3-dimensional search coil system for measuring all 3 rotational degrees of freedom of the eye during VOKR in response to 3-dimensional head

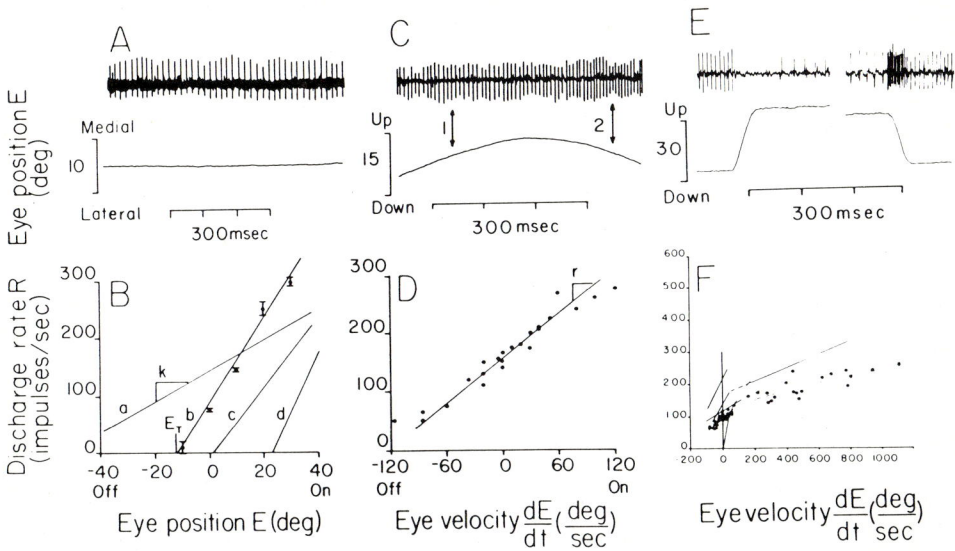

Fig.8  Firing pattern of motoneurons during fixation, slow and rapid
eye movements (from [18])

rotations.

Equation (5), if true, leads to a remarkable simplification of the
integrator problem for the VOKR in 3 dimensions. Consider again Fig.6
for the muscle path on the orbit. Since the part of the muscle from
its insertion at $\underline{r}$ to its departure from the globe at $\underline{s}$ consists
of tendons, which we can assume to be of constant length, the rate
of length change of a non-slackened muscle $i \in \{LR, MR, SR, IR, SO, IO\}$
satisfies (except possibly for the IO)

$$\dot{l}_i = \underline{\dot{s}} \cdot (\underline{o} - \underline{s}) / \|\underline{o} - \underline{s}\| = (\underline{\omega} \wedge \underline{s}) \cdot (\underline{o} - \underline{s}) / \|\underline{o} - \underline{s}\| = r \, \underline{m} \cdot \underline{\omega} \qquad (6)$$

Here $\underline{m} = \underline{s} \wedge \underline{o} / \|\underline{s} \wedge \underline{o}\|$ is the moment unit vector of the muscle, $\underline{\omega}$
is the angular velocity of the eye with radius r. In a compensatory VOKR
$\underline{\omega} = \underline{\omega}_H$ which is satisfied, if $\underline{m}_i \cdot \underline{\omega} = m_i \cdot \underline{\omega}_H$ for all  i . If (5) holds
for the motoneuron  j  of  muscle  i , with constants  u ,  v ,  w  and
if  $\underline{m}_i$ were position independent, then Robinson's 1-dimensional model
of the VOKR [18] would have a simple 3-dimensional generalization:

174

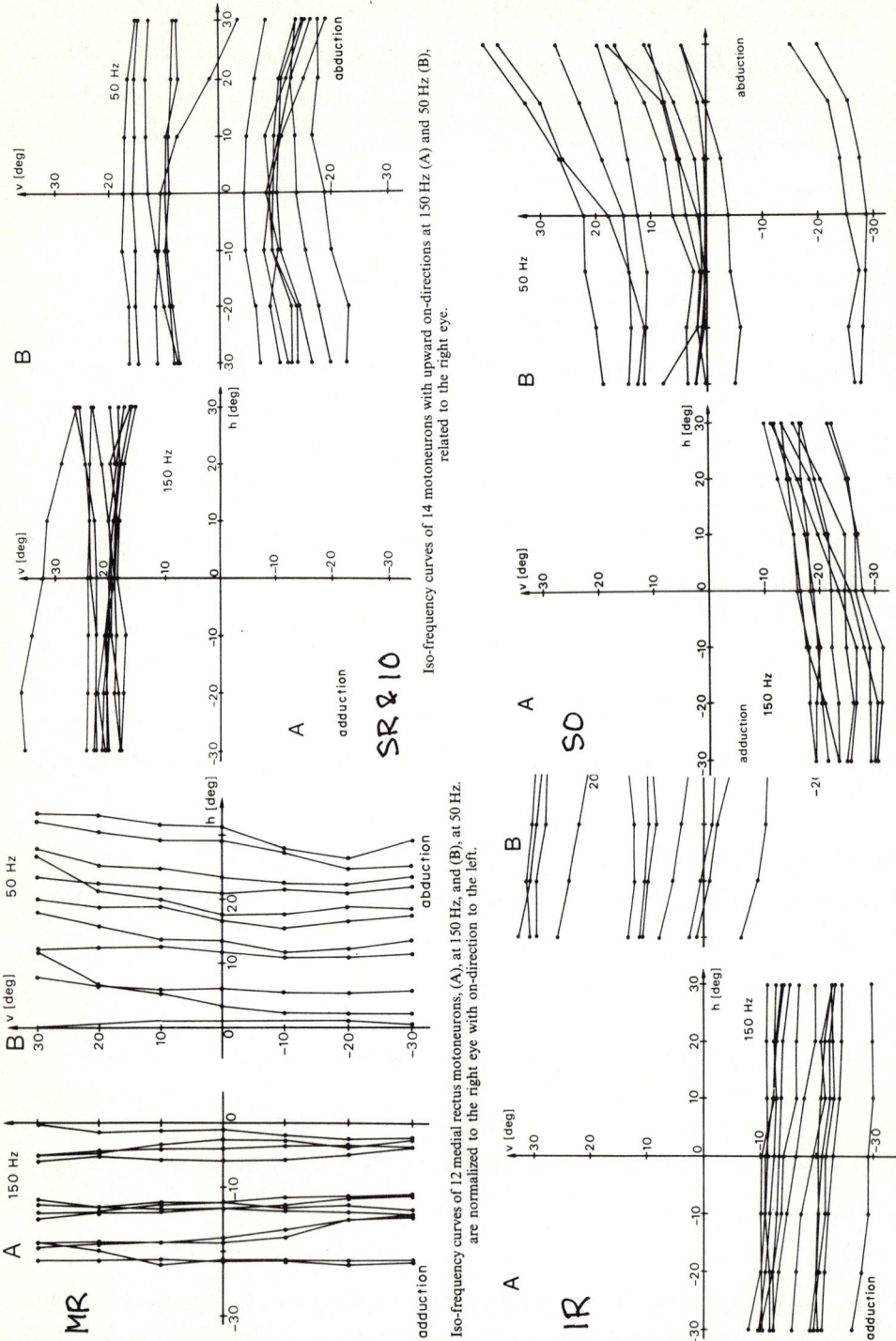

Iso-frequency curves of 14 motoneurons with upward on-directions at 150 Hz (A) and 50 Hz (B), related to the right eye.

Iso-frequency curves of 12 medial rectus motoneurons, (A), at 150 Hz, and (B), at 50 Hz. are normalized to the right eye with on-direction to the left.

Iso-frequency curves of 13 superior oblique motoneurons at 150 Hz (A) and 50 Hz (B), related to the right eye.

Iso-frequency curves of 15 inferior rectus motoneurons at 150 Hz (A) and the right eye.

Fig.9  Iso-frequency curves of motoneurons in the monkey (from [17])

by a direct pathway the fixed linear combination $r$ $w_j$ $m_i \cdot \underline{\omega}_H$ could

could be sent to neuron j, and, through a linear integrator without

position feedback, neuron j would receive its length dependent input.

In reality $\underline{m}_i$ is not constant, and therefore a fully compensatory

VOKR needs a position dependent redistribution of the $\underline{\omega}_H$-proportional

activity in the vestibular nuclei. We have estimated the error in the

VOKR, if the CNS uses instead of $\underline{m}_i(\vartheta,\varphi)$ the combination $\underline{m}_i(0,0)$

appropriate at primary position, by computing for all eye positions ,

satisfying Listing's law the norm $\| M(\vartheta,\varphi) - M(0,0) \|$ $/\|M(0,0)\|$ , where

M has the rows $\underline{m}_{LR} - \underline{m}_{MR}, \underline{m}_{SR} - \underline{m}_{IR}, \underline{m}_{SO} - \underline{m}_{IO}$ and satisfies by (6)

$$d/dt \begin{pmatrix} l_{LR} - l_{MR} \\ l_{SR} - l_{IR} \\ l_{SO} - l_{IO} \end{pmatrix} = r\, M(\vartheta,\varphi)\, \underline{\omega} \tag{7}$$

as first recognized by Robinson [19]. The numerical results in Fig.10

show that an integrator without position feedback generates at most

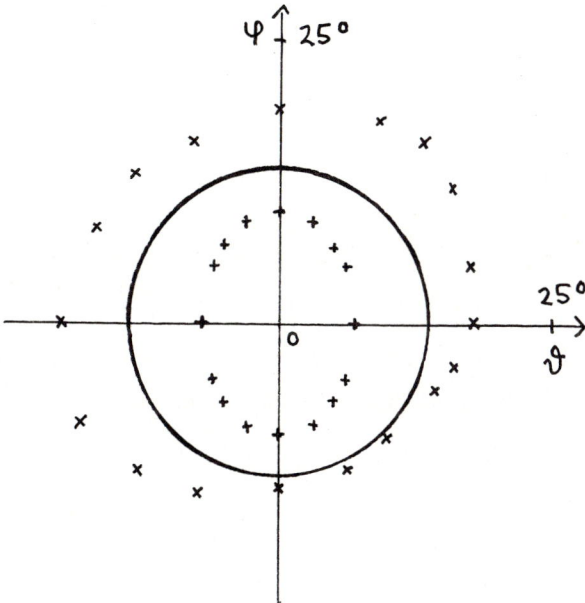

Fig.10 Extent of the periprimary range for the vector approximation
(——) and for a position-independent VOKR integrator in Fick (+++)
and in muscle coordinates (xxx)

10% error in about twice the periprimary range, in which the vector addition law for saccades holds with at most 10% error. This range is considerably larger than the range, where $\| T(\vartheta,\varphi)-T(0,0)\| /\|T(0,0)\|\leq 10\%$. We are working on similar relations for rapid eye movements in 3 dimensions and their to the 2-dimensional visually evoked saccades [20].

In conclusion: By the use of a clever arrangement of the receptor and effector geometry the CNS can use simple direct pathways for important sensorimotor transformations. The range and performance and calibration of the system can then be improved by indirect pathways through the brainstem, cerebellum and cerebral cortex.

Acknowledgement: We are grateful to H.Collewijn, H.Simonsz, J.I.Simpson and T.Vilis for discussions and the communication of their results.

References

[1]   E.R.Kandel "Cellular Basis of Behavior: An Introduction to Behavioral Neurbiology", Freeman, San Francisco, 1976

[2]   E.R.Kandel, J.H.Schwartz "Principles of Neural Sciences", Elsevier, New York, 1985

[3]   F.A.Miles, TINS 7, 303, 1984

[4]   J.I.Simpson, Ann.Rev.Neurosci. 7, 13, 1984

[5]   V.Henn, B.Cohen, L.R.Young, Neurosci.Res.Prog.Bull. 18, 457, 1980

[6]   V.Henn, J.A.Buettner-Ennever, K.Hepp, Human Neurobiol. 1, 77, 1982

[7]   K.Ezure, W.Graf, Neuroscience 12, 85, 1984

[8]   K.Hepp, V.Henn, Springer Series in Synergetics 23, 139, 1983

[9]   E.Mach "Grundlinien der Lehre von den Bewegungsempfindungen", Engelmann, Leipzig, 1875

[10] H.von Helmholtz "Handbuch der Physiologischen Optik", Voss,Leipzig, 1856-1866

[11] D.A.Robinson, Investigative Ophthalmology 14, 801, 1975

[12] L.Ferman, H.Collewijn et al., to appear in Vision Res.

[13] A.T.Bahill, D.Adler, L.Stark, Investigative Ophthalmology, 14, 468, 1975

[14] D.Tweed, T.Vilis, to appear

[15] H.Simonsz, F.Harting, B.J.deWaal, B.Verbeeten, Arch.Ophthalmol 103, 124, 1985

[16] J.I.Simpson, D.Rudinger, H.Reisine, V.Henn, Soc. Neurosci. Abstr. 12, 1986

[17] K.Hepp, V.Henn, Vision Res. 25, 493, 1985

[18] D.A.Robinson "Control of Eye Movements" in "Handbook of Physiology-The Nervous System II", Amer. Physiological Soc., Bethesda, 1981

[19] D.A.Robinson "The coordinates of neurons in the vestibulo-ocular reflex" in "Adaptive Mechanisms in Gaze Control" (A.Berthoz and G.Melvill Jones eds.), Elsevier, Amsterdam, 1985

[20] K.Hepp, T.Vilis, V.Henn, to appear in Annals of the New York Academy of Sciences

## List of Participants

Ueli Aebi, Biozentrum Basel

Dionys Baeriswyl, ETH – Zürich

Jean Pierre Bargetzi, Université de Genève

Chistian Bauer, Universität Zürich

Fran̦cois Bavaud, EPF – Lausanne

Hans Beck, Université de Neuchâtel

Pierre Beran, RCA Zürich

Jakob Bernasconi, Brown Boveri Baden

Marilena Bianchi-Streit, CERN Genève

Hannes Bolterauer, Universität Giessen

Ernesto Bonomi, EPF – Lausanne

Peter Bösiger, ETH – Zürich

Yves Bouligand, CNRS Paris

Eric Bovet, Université de Neuchâtel

Werner Braun, ETH – Zürich

Christoph Bruder, ETH – Zürich

Jean-Pierre Changeux, Institut Pasteur Paris

Bastien Chopard, Université de Genève

Rodney Cotterill, Technical University of Danemark

Yves de Ribaupierre, Université de Lausanne

Pierre Descouts, Université de Genève

Dominique Droz, Prangins

Michel Droz, Université de Genève

Manfred Eigen, Universität Göttingen

Paul Erdös, Université de Lausanne

Fran,cois Fagotto, Université de Neuchâtel

Pierre Fayet, EPF – Lausanne

Sylvia Flesia, Université de Genève

Hans Frauenfelder, University of Illinois

Karl Frei, Basel

Raymond Frésard, Université de Neuchâtel

Ulrike Genz, Universität Konstanz

Ivar Giaever, General Electrics, Schenectady New-York

Claudius Gros, ETH – Zürich

Klaus Hepp, ETH – Zürich

Heinz Horner, Universität Heidelberg

John Hjoth Ipsen, Technical University of Danemark

Hans Jauslin, Université de Genève

Jafferson Kamphorst da Silva, Université de Genève

Werner Känzig, ETH – Zürich

Wolfgang Kinzel, Universität Giessen

Otto Krisement, Universität Münster

Murat Kunt, EPF – Lausanne

Hans Rudolf Lüscher, Universität Zürich

Daniel Maeder, Université de Genève

Andreas Malaspinas, Université de Genève

Pierro Martinoli, Université de Neuchâtel

Ole Mouritsen, Technical University of Danemark

Jean-Pierre Nadal, ENS Paris

Ernest Niebur, Université de Lausanne

Martin Peter, Université de Genève

Serge Poitry, Université de Genève

John Ross, Stanford University

Fran,cois Rothen, Université de Lausanne

Daniel Saint-James, Collège de France Paris

Maria Maddalena Sperotto, University of Copenhagen

Samuel Steinemann, Université de Lausanne

Humbert Suarez, Université de Genève

André Vallat, Université de Neuchâtel

Herbert Wagner, Universität München

Helene Widmer, Hôpital Cantonal de Genève

Hans Jürg Wiesmann, Brown Boveri Baden

Hans Rudolf Zeller, Brown Boveri Baden

Fu-chu Zhang, ETH – Zürich

Martin Zulauf, Basel

# Lecture Notes in Physics

G. Maret, J. Kiepenheuer, N. Boccara (Eds.)

# Biophysical Effects of Steady Magnetic Fields

Proceedings of the Workshop, Les Houches, France,
February 26–March 5, 1986

1986. 129 figures. XII, 231 pages. (Springer Proceedings in
Physics, Volume 11). ISBN 3-540-16992-X

**Contents:** Diamagnetic Orientation of Macromolecules and
Membranes. – Ferrofluids. – Biological and Chemical Photo-
reactions. – Physiological Effects and Animal Development. –
Animal Orientation and Magnetoreception. – Applications in
NMR and Medicine. – Index of Contributors.


M. E. Michel-Beyerle (Ed.)

# Antennas and Reaction Centers of Photosynthetic Bacteria

**Structure, Interactions and Dynamics**

Proceedings of an International Workshop Feldafing,
Bavaria, F. R. G., March 23–25, 1985

1985. 168 figures. XI, 367 pages. (Springer Series in Chemical
Physics, Volume 42). ISBN 3-540-16154-6

**Contents:** Antennas: Structure and Energy Transfer. – Reaction Centers: Structure and Interactions. – Electron-Transfer:
Theory and Model Systems. – Reaction Centers: Structure
and Dynamics. – Model Systems on Structure of Antennas
and Reaction Centers. – Index of Contributors.


T. W. Barrett, H. A. Pohl (Eds.)

# Energy Transfer Dynamics

**Studies and Essays in Honor of Herbert Fröhlich on His
Eightieth Birthday**

1987. 87 figures. XIII, 352 pages. ISBN 3-540-17502-4

Springer-Verlag
Berlin Heidelberg New York
London Paris Tokyo

Springer